MAGNETISM AND MAGNETIC MATERIALS

Magnetism and Magnetic Materials

Second Edition

J. P. JAKUBOVICS

*University Lecturer in Metallurgy
and Science of Materials*

The University of Oxford

THE INSTITUTE OF MATERIALS

Book 573
First published in 1994 by
The Institute of Materials
1 Carlton House Terrace
London SW1Y 5DB

British Library Cataloguing-in-Publication Data

Jakubovics, J. P.
Magnetism and Magnetic Materials. –
2Rev. ed
I. Title
538.8

ISBN 0-901716-54-5

Reproduced from the author's camera-copy
Halftone origination by the University Press, Cambridge
Printed and bound in Great Britain at
The University Press, Cambridge

Contents

Preface

THE contents of this book started life as a course of eight lectures given to third year undergraduates reading Metallurgy and Science of Materials at Oxford. However, the book differs from the lecture course in two respects.

Firstly, I have tried to explain everything from first principles as far as possible, in the hope that readers new to the subject, including those with relatively little knowledge of mathematics, can benefit from it. Secondly, the fact that the course is limited to eight lectures imposes a restriction on the amount of material that can be covered, and therefore I have tried to include details for which there is no time in the lecture course.

This book is not intended to be a self-contained source of information—there are many excellent texts that serve that purpose already. It should however enable readers to gain an understanding of a subject that may be new to them, or one they may have found difficult before, and it should encourage them to approach more advanced texts without feeling intimidated. For that reason I have included some advice on further reading in an appendix. At the same time, I hope that readers with some knowledge of magnetism (I trust my students can be included in that category) may find some useful information in the book.

Readers who explore the subject further will find not only many more details on the topics dealt with in this book, but entirely new areas that had to be excluded because this was meant to be a small volume. Examples of the latter are semihard materials, rare earth-transition metal compounds with giant magnetostriction, and ferromagnetic liquids. Information on these topics and many others will be found in the books recommended in the appendix.

It is a pleasure to thank present and former colleagues and other friends for supplying figures for the book. Fig. 3.21 was contributed by G. A. Jones, Fig. 3.23 by A. Hubert, Fig. 3.24 by A. Hubert and M. Rührig, Fig. 3.28 by M. R. Scheinfein, Fig. 3.29 by I. McFadyen and Fig. 3.30 by J. E. A. Miltat. Fig. 3.25 was taken by S. J. Newnham, Fig. 3.26 by D. C. Joy and Fig. 3.27 by D. J. Fathers. I am grateful to them all for their contributions. I am also grateful to S. M. Baker for assistance with obtaining Fig. 3.22 and the cover photograph, and to A. K. Petford-Long for reading parts of the manuscript. I also thank Oxford University Computing Service, and R. L. Hutchings in particular, for their help with typesetting this book.

Oxford   J. P. J.
1994

1

Fundamentals of magnetism

1.1 What is a magnetic material?

MAGNETIC phenomena have been known and exploited for many centuries. The earliest experiences with magnetism involved magnetite, the only material that occurs naturally in a magnetic state. This mineral was also known as lodestone, after its property of aligning itself in certain directions if allowed to rotate freely, thus being able to indicate the positions of north and south, and to some extent also latitude. The other well-known property of lodestone is that two pieces of it can attract or even repel each other.

After the production of iron from its ores had become possible, it was realised that magnetite could also attract iron. There are very many magnetic materials known to-day, and it is therefore useful first of all to give a very empirical rule for what might be called a *magnetic material*.

If two objects attract each other and also repel each other (depending on their relative orientations), then these objects might be called magnets. There are also other objects that are attracted to, but not repelled by magnets, and are not attracted or repelled by each other. Such objects may be said to consist of magnetic materials. The majority of materials are non-magnetic by this definition.

This simple definition, while it may help to relate magnetism to phenomena from everyday life, is not really very satisfactory. The first reason for this is that the magnetic state of a material is not constant, but may be changed in various ways. Thus, magnetite is not always a *magnet*, although (at least at room temperature) it is always a *magnetic material*. It will also turn out that all materials are attracted or repelled by magnets, but in most cases, the forces involved are extremely small. The second reason why a more precise definition is needed is that there is another type of force that can be exerted by one material on another, which does not arise from magnetism. There are many examples of materials that can acquire the property of being able to attract or repel small objects after being rubbed together with another material. This kind of force, which is called electrostatic, has also been known for many centuries.

1

1.2   The relationship between electricity and magnetism

Electrostatic and magnetic forces differ from each other in one very important respect. The electrostatic force between two bodies may be purely repulsive, such that the two bodies always repel each other regardless of their relative orientations. With two magnets, this can never be the case: even if they repel each other in some particular relative orientation, it is possible to find another orientation in which they attract. This has led to the assumption that electrostatic forces were due to electric charges, which may be of two kinds, like charges repelling and opposite charges attracting each other. In the case of magnetism, even if 'magnetic charges' exist, they always behave as if they came in pairs of equal magnitude and opposite sign, a small distance away from each other. A pair of charges of this type is called a *dipole*. No one has so far succeeded in separating magnetic dipoles into individual charges.

During the eighteenth century, the electric cell was developed, and it became possible to maintain a continuous flow of electric charges, called currents, through conductors. This development led to experiments of fundamental importance during the first half of the nineteenth century, which established the connection between electricity and magnetism. It was first demonstrated that circuits that had currents flowing through them behaved in similar ways to magnets. A freely rotating magnetic dipole, such as a compass needle, could be deflected if placed near a current-carrying wire. If a loop of wire carrying a current is free to rotate, it will assume a certain geographical orientation. When two current-carrying wires are placed near each other, they exert a force on each other. The magnitude of the force is proportional to the product of the currents flowing in the two wires. Later, it became evident that it was also possible for magnetic phenomena to produce electrical forces. The important point, discovered by Michael Faraday, was that electrical forces could only be produced while something was changing with time: the magnitude of the current producing the magnetic effect, or the positions of circuits relative to each other, or the position of a magnet.

For our discussion of magnetic materials, we shall have to be able to compare the magnitudes of certain quantities, and we now have to define them. In order to be able to measure magnetic forces, we might use a small compass needle attached to a fixed point by a torsion suspension. If we placed such an instrument inside a coil of wire carrying a current, we would find that the compass needle tended to align itself parallel to the axis of the coil. If the coil is long compared with the diameter of its cross-section, then the aligning force, or couple, acting on the compass needle has the same magnitude at all points inside the coil (at least at points not too near the ends of the coil). For a given compass needle, the couple is proportional to the current, i, flowing in the wire and to the number of turns of wire per unit length of coil, n. A region of space where a compass needle is acted on by a couple is said to be permeated by a *magnetic field*. The magnitude of this field is proportional to the couple acting on a given compass needle. For the coil, the magnetic field is proportional to i and to n, and we define the field, H, to be

$$H = ni. \tag{1.1}$$

Since i is measured in ampères and n in number of turns per metre, H can be expressed in ampères per metre or $A\,m^{-1}$. Apart from H, the couple, or torque, depends also on the nature of the compass needle and on its orientation relative to the axis of the coil, being a maximum when the needle is perpendicular to the axis, and zero when it is parallel. It is important to realise therefore that the magnetic field has not only a magnitude, but also a direction, just as any force has a direction as well as a magnitude. The magnetic field of a coil is parallel to the axis of the coil.

We also need a way to compare the strength of different compass needles. As we shall see, this will present us with a problem. It will help if we first consider the behaviour of electric charges in an electric field. An electric field can be produced between two large parallel metal plates, each connected to one of the terminals of a battery. We could in principle make an electric compass out of a thin rod of insulating material such as a short piece of glass fibre, by placing a positive charge on one end and an equal amount of negative charge on the other. Such an electric compass would experience a couple in the electric field, just as the magnetic compass does in the magnetic field. But in the case of the electric compass, it can be shown that the couple is proportional to the magnitude of the positive (or the negative) charge, and to the distance between these charges. A pair of opposite charges separated by a distance is called an *electric dipole*, and the magnitude of the dipole is defined to be the product of the amount of positive charge and its separation from the negative charge.

In the case of the electric compass, we can verify that it consists of two opposite charges—we can cut the compass in half, and we will have two separate pieces carrying opposite charges. In the electric field, the two pieces experience forces in opposite directions to each other. The difficulty with the magnetic compass is that if we cut it in half, we do not end up with two pieces that carry magnetic charges of opposite sign. The two pieces still only experience a couple, not a net force, in the magnetic field (provided the field is uniform). We can continue to cut the compass into more and more pieces, but we will just keep producing smaller and smaller dipoles— the dipoles can never be separated into individual charges. It does not therefore make sense to define a magnetic dipole as the product of a charge and a distance, because the charges do not lead a separate existence. We can still compare the magnitude of two dipoles by the relative magnitude of the couple they experience when placed in the same magnetic field, at right angles to it. We can call the magnitude of the dipole the *dipole moment*, denoted by m. Both electric dipoles and magnetic dipoles have directions as well as magnitudes. In the case of an electric dipole, we can define the direction as pointing from the negative towards the positive charge. For the magnetic dipole, we cannot use this definition. However, another way to specify the direction is to find the orientation in which the couple acting on the dipole is a maximum. The dipole is then perpendicular both to the field and to the axis about which it is trying to rotate.

A plane loop of wire carrying a current also experiences a couple in a magnetic field. It is found that the couple is proportional to the current, i, and to the area of the loop, S. The loop therefore has a magnetic dipole moment of magnitude

$$m = iS. \tag{1.2}$$

It can be shown by experiment that the dipole is perpendicular to the plane of the loop.

As the torque, G, acting on a dipole m in a magnetic field H is proportional to H and to m, it may seem convenient to define m as being of unit magnitude if it experiences a unit torque (1 newton metre or $1\,\mathrm{N\,m}$) in a magnetic field of $1\,\mathrm{A\,m^{-1}}$. However, in the system of units widely adopted nowadays, the SI system, the torque on a unit dipole in a unit field is not equal to $1\,\mathrm{N\,m}$. We express the torque as

$$G = \mu_0 m H \sin\theta, \tag{1.3}$$

where μ_0 is a universal constant called the *permeability of vacuum*, and θ is the angle between the magnetic field and the dipole. In the SI system, μ_0 is defined as $4\pi \times 10^{-7}$ henry per metre ($\mathrm{H\,m^{-1}}$).

If the space inside the coil is filled with some material rather than vacuum, the torque acting on the dipole will be different, depending on the material. It is necessary to define another kind of magnetic field, which determines the torque acting on the dipole in the general case when the dipole is surrounded with a material. This second type of magnetic field is called the *magnetic induction*, denoted by B. Eq. (1.3) now becomes

$$G = mB \sin\theta, \tag{1.4}$$

which is more generally valid than eq. (1.3). The unit of B is the tesla (T). Eq. (1.4) shows that the torque on a dipole of moment $1\,\mathrm{A\,m^2}$ in a magnetic induction of $1\,\mathrm{T}$ is $1\,\mathrm{N\,m}$ if m is perpendicular to B. (Readers familiar with vector notation will realise that eqs (1.3) and (1.4) can be written in a more compact form. A fuller explanation is given in appendix C.)

1.3 The effect of magnetic fields on materials

Eqs (1.3) and (1.4) show that in a vacuum, B and H are related by

$$B = \mu_0 H. \tag{1.5}$$

B and H are always parallel to each other in a vacuum. If a material is present, eq. (1.5) does not hold. The material acquires a dipole moment in the presence of the magnetic field. The way in which this dipole moment arises will be dealt with in subsequent chapters. For now, it is sufficient to know how this dipole moment can be expressed quantitatively. Obviously, the magnitude of the dipole acquired by a specimen of a given material in a given field is proportional to its volume. We can therefore define the magnetization, M, of the material as being its dipole moment per unit volume. The magnetization is related to B and H by

$$B = \mu_0(H + M), \tag{1.6}$$

which is a generalisation of eq. (1.5). Eq. (1.6) shows that M is measured in the same units as H, i.e. in $\mathrm{A\,m^{-1}}$.

The magnetic induction, B, is particularly important when we consider electrical forces produced by magnetic phenomena, mentioned above. Consider a wire loop of area S, with magnetic induction B passing through it. If B is perpendicular to the plane of the loop, we say that there is a magnetic flux, Φ, passing through the loop, given by

$$\Phi = BS. \tag{1.7}$$

If B is not perpendicular to the loop, then the flux is determined by the area we would see if we were looking at the loop along the direction of B. Expressed mathematically, eq. (1.7) then becomes

$$\Phi = BS \cos\theta, \tag{1.8}$$

where θ is the angle between the direction of B and the perpendicular to the plane of the loop. Flux is measured in webers (Wb), and therefore an induction of $1\,\mathrm{T}$ can also be written as $1\,\mathrm{Wb\,m^{-2}}$. As we can see, the magnetic induction can also be thought of as the density of flux (flux per unit area), and is sometimes referred to as that.

The law discovered by Faraday states that the force driving the current round the circuit, i.e. the electromotive force (measured in volts) is equal to the rate of change of flux through the circuit. As discussed below, this makes magnetic materials important in the production of electricity. Before outlining briefly the applications of magnetic materials, we first need to consider some of the properties used for characterising them.

1.4 The most important magnetic property

The magnetization of a material in general depends on the magnetic field acting on it. For many materials, M is proportional to H (at least when H is not too large) and we may write

$$M = \chi H, \tag{1.9}$$

5

where χ, the *magnetic susceptibility*, is a property of the material. Since M and H have the same dimensions, χ is dimensionless. We can combine eq. (1.9) with eq. (1.6), to get

$$B = \mu_0(1 + \chi)H. \tag{1.10}$$

Thus $1 + \chi$ expresses the proportionality of B and H. We define

$$\mu = 1 + \chi, \tag{1.11}$$

where μ is the *magnetic permeability*. Eqs (1.10) and (1.11) give

$$B = \mu_0\mu H. \tag{1.12}$$

Either μ or χ may be used to characterise a material. As we shall see in the next few chapters, χ ranges from values close to 0, both positive and negative, to positive values much greater than 1, for different materials. For materials that have very small susceptibilities, it is much more convenient to use χ than μ. (If, for example, $\chi = 10^{-6}$, μ would be 1.000001 and if $\chi = -10^{-6}$, μ would be 0.999999.) Since such materials are chiefly of interest to physicists, they tend to regard χ as the fundamental magnetic property. On the other hand, engineers are usually interested in materials with large values of χ and of μ. For engineers, it is of more interest to know the magnetic induction produced by a material than its magnetization. Materials of practical interest are therefore usually characterised by their permeabilities.

1.5   Applications of magnetic materials

We end this introductory chapter with some general remarks about the chief uses of magnetic materials. There are three types of applications for which they are used in large quantities.

Firstly, as we have seen, magnetic fields can be produced by currents passing through coils. By using magnetic materials as the cores of the coils, the fields produced can be enhanced, or they can be concentrated into small volumes of space. The first application of magnetic materials is therefore as electromagnet cores.

Secondly, magnetic materials can aid the transfer of electrical energy from one circuit to another. This is usually done by inserting a coil into each circuit and bringing the two coils into close proximity with each other. If a varying (e.g. alternating) current is passed through one circuit, a varying flux will be produced, and according to Faraday's law, a current will be induced in the second circuit. The magnetic induction, and hence the flux passing through the second circuit, can be enhanced by magnetic materials.

They are therefore used as transformer cores. The two applications are related to some extent, since in both cases, magnetic materials are used to amplify the field, or magnetic induction, produced by electric currents. However, in the first case, the fields are usually required to be constant, whereas in the second case, they must vary with time. The properties required of the materials for the two applications are therefore not the same.

The third main application is for the production of magnetic fields without electric currents. This can be achieved by the use of permanent magnets. Materials for this type of application must have quite different properties from those used for the first two applications.

There are other, more specialised applications for magnetic materials. Some of these applications can be regarded as special cases of one of the three categories above. More details will be given in later chapters.

2

Classification of materials by magnetic properties

2.1 Criteria for classifying materials

2.1.1 Magnetic susceptibility and its dependence on temperature

As we saw in chapter 1, the most important magnetic property of a material is its susceptibility, χ. Since our main interest lies in materials that have practical applications, it would be tempting to begin our classification by dividing materials into two categories: those with $|\chi| \ll 1$, which do not have useful magnetic properties, and those with $\chi \gg 1$, which could potentially be useful. However, if we wish to gain a deeper understanding of magnetic materials, it is better to start by separating materials for which χ is negative from those for which χ is positive. As we shall see, materials with large values of χ can be regarded as rather special, exceptional cases in the latter category. The importance of this classification becomes evident when we consider the variation of χ with temperature. In materials where the susceptibility is negative, it nearly always has a constant value independent of temperature. In all other materials, the susceptibility varies with temperature. At high temperatures at least, χ shows a decreasing trend with increasing temperature, T, in all materials.

Before discussing the behaviour of the susceptibility in more detail, we have to qualify the above statement concerning the temperature variation of the susceptibility. We have defined χ as the ratio of the magnetization, i.e. magnetic moment per unit volume, to the applied field. However, when the temperature changes, it affects the volume of the material as well as its magnetic moment, because of thermal expansion. Our discussion of the temperature dependence of χ is really only valid for an idealised material in which thermal expansion does not take place. Experimentally, it is much easier to measure accurately the mass of a specimen than its volume. It is customary therefore to express the results of measurements in terms of magnetic moment per unit mass. The ratio of this quantity and the applied field is called the mass susceptibility, χ_m. The relationship between χ and χ_m is

$$\chi = \rho \chi_m, \tag{2.1}$$

9

where ρ is the density. Therefore χ_m has to be expressed in units of $m^3 \, kg^{-1}$. Although in real materials, only χ_m is unaffected by thermal expansion, it is still more convenient to use χ in general discussions of magnetic properties, since χ is dimensionless. It will have to be remembered however that in our discussions, we are neglecting thermal expansion.

Within the class of materials having positive susceptibility, the magnitude of the susceptibility varies over a very wide range—from values several orders of magnitude less than one to values several orders of magnitude greater than one. Again, it would not be very useful to subdivide this class simply according to the magnitude of the susceptibility. We have already seen that at sufficiently high temperatures, the susceptibility decreases with increasing temperature for all materials in this class. Experimentally, it is found that all these materials follow the relationship

$$\chi = \frac{C}{T \pm \theta} \tag{2.2}$$

more or less exactly, for sufficiently high T. In the equation, C and θ are positive constants, different for each material.

In order to carry the subdivision further, we have to consider the behaviour of χ as T decreases. In some materials, it is found that $\theta = 0$ and that eq. (2.2) is obeyed down to the lowest temperatures at which measurements have been made. These are the simplest materials in the class, and are called *paramagnetic*. Materials in the other main class, for which χ is negative, are called *diamagnetic*.

2.1.2 Critical temperature and spontaneous magnetization

In all other materials, eq. (2.2) breaks down as T decreases. They all have a *critical temperature* below which the variation of χ with T is very different from its variation above. The simplest case to consider is the one where the negative sign occurs in eq. (2.2). Then, obviously, χ becomes very large when T approaches θ. At $T = \theta$, χ would be infinite. An infinite susceptibility means that a finite magnetization can exist even in zero applied field. This must obviously happen in permanent magnets. There are other materials that become strongly magnetized in relatively small fields, but do not retain a significant magnetization when the field is removed. (Even 'permanent' magnets can become demagnetized!) Thus, the magnetization of a material in zero field can have a range of different values and therefore cannot be regarded as a property of the material. However, it is found that if a relatively small field is applied to these materials, the magnetization tends to a constant value, called the *saturation magnetization*, M_s, and that the magnetization does not increase significantly with further increases in applied field. For reasons that will become apparent later, we should equate the 'spontaneous magnetization' with M_s rather than with the magnetization actually observed in zero field. M_s is a function of temperature, becoming zero at the critical temperature.

The class of materials for which θ has a negative sign in eq. (2.2) is further subdivided into two groups. In one group, eq. (2.2) is followed to a reasonable accuracy at all temperatures above $T = \theta$, with small departures occurring mainly around $T = \theta$. The critical temperature is approximately equal to θ, and is called the *Curie temperature*, θ_C. Below θ_C, M_s varies with temperature in a similar way for all materials in the group. Of course, θ_C is different for each material, and M_s at a given temperature is also different. But if we measure M_s for two different materials at temperatures such that T/θ_C is the same for the two materials, and we also measure M_s at a very low temperature for each, we will find that they have the same value of $M_s(T)/M_s(0)$. Here $M_s(T)$ denotes the value of M_s at the absolute temperature T. In other words, if we plot a graph of $M_s(T)/M_s(0)$ against T/θ_C for different materials in this group, the plots will be approximately superimposed. Fig. 2.1 shows this behaviour schematically. It is seen that the spontaneous magnetization tends to a constant value as T tends to absolute zero. As T increases, M_s decreases more and more rapidly. Materials in this group are called *ferromagnetic*.

In materials of the other subgroup, called *ferrimagnetic* materials, significant departures from eq. (2.2) occur over a range of temperatures. The equation is only followed at temperatures large compared with the Curie temperature. For ferrimagnetic materials, plots of $M_s(T)/M_s(0)$ against T/θ_C for different materials cannot usually be superimposed. A more detailed discussion of ferrimagnetic materials can be left for a later chapter.

The final group of materials we need to discuss resemble the paramagnetic group in that they have small, positive susceptibilities at all temperatures. However, their susceptibilities do not increase steadily as the temperature decreases all the way to absolute zero. At high temperatures, they follow eq. (2.2), with θ usually having a positive sign. At a critical temperature, in this case called the *Néel temperature*, θ_N, they cease to obey eq. (2.2). But below θ_N, the susceptibility generally decreases with decreasing temperature. These materials are called *antiferromagnetic*.

To complete the classification, we should also mention another class of materials whose properties resemble those of antiferromagnetic materials in many ways. Their susceptibilities also increase as T decreases from a large value. They also go through some kind of transition and at low temperatures, they do not obey eq. (2.2). However, the nature of the transition and the low-temperature behaviour are more complicated than in antiferromagnetic materials. These materials are called *spin glasses*, and they are currently of greater interest to physicists, because their properties are not fully understood.

We shall also see later that there are some materials that do not fall clearly into one or other of the above categories. There are, for example, materials that exhibit either ferromagnetism or antiferromagnetism at

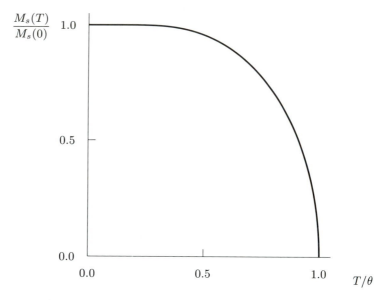

FIG. 2.1   Variation of spontaneous magnetization with temperature in ferromagnetic materials.

different temperatures. However, for the moment, we will concentrate on
simple, clear-cut cases.

The five main classes of materials and the criteria by which they are
distinguished from each other are summarised in Table 2.1. The properties
we have discussed so far are shown in columns 2–5 of the table.

We will now discuss whether it is possible to predict which class a material belongs to if we know its chemical composition and, if it is a solid,
its crystal structure. The information needed for this is given in column 6
of Table 2.1.

2.1.3   The magnetism of atoms

We have already seen that magnetism is associated with electric currents,
i.e. the motion of electric charges. According to the classical model of the
atom, the electrons move in orbits round the nucleus. There is therefore a
magnetic moment associated with each orbiting electron. This model gives
us some idea of how the atom responds to an applied magnetic field. We
cannot expect more than a qualitative agreement with the behaviour of
real materials. To get reliable results, we should use the quantum model of
the atom, which has long since superseded the classical theory. However,
for the moment, we will sacrifice accuracy for simplicity of insight.

As a simple example, we consider a helium atom. There are two electrons orbiting the atom, and each orbit should therefore have a magnetic moment. However, the electrons may be thought of as orbiting in opposite senses, so that the two magnetic moments are in opposite directions and cancel each other out. Therefore, when there is no applied magnetic field, the helium atom has no magnetic moment. If a magnetic field is applied, the effect on each orbiting electron is the same as it would be on any electrical circuit. According to Faraday's law, an electromotive force will be induced in each circuit and the current will change. In a macroscopic electrical circuit, there is always some resistance present, so that any induced current will stop flowing very rapidly as soon as the magnetic field has stopped varying. However, in the atom, there is no resistance to the motion of the electrons, and the induced current will therefore persist indefinitely. Only when the applied field is removed, will there be an equal and opposite current induced which just cancels out the current induced initially. Hence, in this case, the induced current is proportional to the magnetic flux through the electron orbit, rather than to the rate of change of the flux.

The induced current gives rise to an extra magnetic moment in each orbit. However, according to the laws of electromagnetism (the relevant law is usually called *Lenz's law*), the resulting magnetic moment is in the opposite direction to the field that has induced the current. Let us assume that the original magnetic moment of one of the orbits was parallel to the applied field, and that of the other orbit was antiparallel. Then the effect of the field is to decrease the first moment and increase the second slightly. The atom now has a net moment which is in the opposite direction to the field. The magnitude of the moment is proportional to the field. Since each atom develops a moment when the field is applied, helium has a negative susceptibility, i.e. it is diamagnetic.

This simple model of diamagnetism would even enable us to estimate the magnitude of χ, and our estimate would not be very far from the measured value—perhaps within a factor of two from it. Nevertheless, our model of the atom is very unrealistic. The electronic orbits in helium do not have a magnetic moment at all in zero field. If we insist on a classical analogy, perhaps we should think of the electrons in helium not as moving in circular orbits, but as oscillating along a line through the nucleus (thus having no magnetic moment), with the orientation of the line being indeterminate, or constantly rotating. This makes no difference however: even if there is no moment initially, an applied field can induce the appropriate moment in the opposite direction to itself, in each of the electronic orbits. There is another important modification we must make to our model of the atom. Magnetic moments arise not only from the orbital motion of the electrons. Each electron behaves as if it had a magnetic moment of its own. The classical explanation of this moment would have to be based on the assumption that the charge of the electron is not concentrated at a point, but is slightly spread out, and this charge distribution is constantly spinning on its own axis. Hence, the source of this magnetic moment is referred to as the *spin*

Classification of materials by magnetic properties

Class	Critical temperature	Magnitude of χ	Temperature variation of χ
Diamagnetic	None	Approximately -10^{-6} to -10^{-5}	Constant
Paramagnetic	None	Approximately $+10^{-5}$ to $+10^{-3}$	$\chi = C/T$
Ferromagnetic	Curie temperature, θ_C	Large (below θ_C)	Above θ_C, $\chi = C/(T - \theta)$, with $\theta \approx \theta_C$
Antiferromagnetic	Néel temperature, θ_N	As paramagnetic	Above θ_N, $\chi = C/(T \pm \theta)$, with $\theta \neq \theta_N$; below θ_N, χ decreases, anisotropic
Ferrimagnetic	Curie temperature, θ_C	As ferromagnetic	Above θ_C, $\chi \approx C/(T \pm \theta)$, with $\theta \neq \theta_C$

Table 2.1   Classification of materials

Spontaneous magnetization	Structure on atomic scale	Examples
None	Atoms have no permanent dipole moments.	Inert gases; many metals, e.g. Cu, Hg, Bi; non-metallic elements, e.g. B, Si, P, S; many ions, e.g. Na^+, Cl^-, and their salts; most diatomic molecules, e.g. H_2, N_2; water; most organic compounds.
None	Atoms have permanent dipole moments. Neighbouring moments do not interact.	Some metals, e.g. Cr, Mn; some diatomic gases, e.g. O_2, NO; ions of transition metals and rare earth metals, and their salts; rare earth oxides.
Below θ_C, $M_s(T)/M_s(0)$ against T/θ_C follows a universal curve; above θ_C, none	Atoms have permanent dipole moments. Interaction produces $\uparrow\uparrow$ alignment.	Transition metals Fe,H Co, Ni; rare earths with $64 \leq Z \leq 69$; alloys of ferromagnetic elements; some alloys of Mn, e.g. MnBi, Cu_2MnAl.
None	Atoms have permanent dipole moments. Interaction produces $\uparrow\downarrow$ alignment.	Many compounds of transition metals, e.g. MnO, CoO, NiO, Cr_2O_3, MnS, MnSe, $CuCl_2$.
Below θ_C, does not follow universal curve; above θ_C, none	Atoms have permanent dipole moments. Interaction produces $\uparrow\downarrow$ alignment, but moments are unequal.	Fe_3O_4 (magnetite); γ-Fe_2O_3 (maghemite); mixed oxides of iron and other elements.

according to magnetic properties

of the electron. The spin magnetic moment of the electron has a constant magnitude, and in a magnetic field, it can only orient itself either parallel or antiparallel to the field. These two facts by themselves demonstrate that the spin can only be explained satisfactorily on the basis of a quantum mechanical model.

In the helium atom, the spin magnetic moments of the two electrons are in opposite directions, and so they do not contribute to the magnetization. This model therefore enables us to explain the diamagnetism of helium.

Now we consider a different case, that of the triply charged ion of iron, the ferric ion (Fe^{3+}). It has 23 electrons. Two of them are in the innermost orbits, similar to those in helium, and therefore they have no net magnetic moment. The next eight make up the second electron 'shell', and again, we can pair off the electrons in the shell so that the net moment of each shell is zero. This is in fact the case for any electron shell that is closed, i.e. has the maximum permissible number of electrons in it. It is also the case for the third electron shell in Fe^{3+}, containing electrons nos. 11 to 18. This leaves 5 electrons that cannot all be paired off. It turns out that the magnetic moments arising from the orbital motion of these electrons do cancel out, but that the spin moments of all the electrons are parallel to one another. The ion therefore has a net magnetic moment. Atomic magnetic moments are measured in a unit called the *Bohr magneton*; it has a magnitude of $\approx 9.27 \times 10^{-24}\,A\,m^2$, (see appendix A) and is denoted by μ_B. Each spin has a moment of $1\mu_B$, and the Fe^{3+} ion therefore has a net moment of $5\mu_B$.

When a magnetic field is applied to the Fe^{3+} ion, the magnetic moment tries to rotate into an orientation parallel to the field. If it succeeds to rotate even by a small amount, the magnetization of a specimen containing Fe^{3+} ions will increase. Hence the susceptibility is expected to be positive. The extent to which the magnetic moment succeeds in rotating towards the field will have to be discussed more fully later.

As in helium, the magnetic field also induces a negative magnetic moment in the electron orbits of the Fe^{3+} ion. However, this negative moment is much smaller than the $5\mu_B$ already present. It is much smaller even than the change of moment caused by the reorientation of the $5\mu_B$. Hence, it may be concluded that materials in which none of the atoms, or molecules, have a permanent magnetic dipole moment are diamagnetic, whereas materials that contain some atoms or molecules with a permanent magnetic moment are paramagnetic, ferromagnetic, antiferromagnetic or ferrimagnetic.

It is much more difficult to predict which of the latter four categories a material belongs to, once we know that it contains permanent magnetic moments. However, using some hindsight, we can state that the distinction between paramagnetic materials and materials in the other three classes is that in the former, the magnetic moments point in random directions at all temperatures in zero magnetic field, whereas in the latter, the magnetic

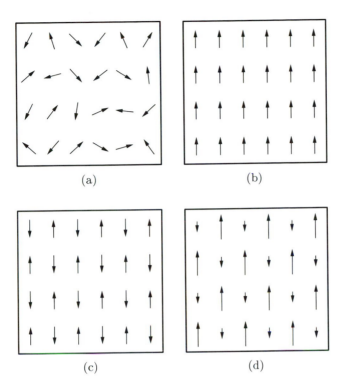

FIG. 2.2 Schematic diagrams of the alignment of magnetic moments, (a) in paramagnetic materials at all temperatures, and (b) in ferromagnetic materials, (c) in antiferromagnetic materials, and (d) in ferrimagnetic materials at low temperatures.

moments of the atoms tend to align themselves with those of their neighbours. In a ferromagnetic material, the moments try to become parallel to one another, at least on a microscopic scale. In an antiferromagnetic material, the tendency is for neighbouring moments to be aligned antiparallel to each other. In a ferrimagnetic material, the tendency is also for antiparallel alignment, but either the number or the size of the moments in one direction is greater than in the other direction. A schematic diagram of the moments in each type of material is shown in Fig. 2.2. The alignment of moments shown in Fig. 2.2(b), (c) and (d) is present only below the critical temperature. Above this temperature, the atomic moments in all four types of materials are randomly oriented as in Fig. 2.2(a).

The alignment of moments can be investigated experimentally by neutron diffraction. A great many materials have been studied by this technique, and the types of alignment have been found to correspond to what

17

was expected from measurements of magnetization and susceptibility.

2.1.4   Examples of materials of different types

We will not go into details about the reasons for the alignment of moments in various materials. Rather than trying to answer the question of why a given material should be paramagnetic, ferromagnetic, etc., we will end this section with some examples of materials in each category. The examples are listed in the last column of Table 2.1.

Diamagnetism is relatively easy to predict. As expected, all noble gases, such as helium, neon, argon, etc., are diamagnetic. So are nearly all materials consisting of diatomic molecules, such as hydrogen, nitrogen, carbon monoxide and hydrogen chloride. (However, oxygen and nitrous oxide are exceptions—they are paramagnetic.) Also diamagnetic are materials consisting of ions with noble gas type electron configurations, such as sodium chloride. Non-metallic elements such as boron,indexBoron silicon and phosphorus are all diamagnetic, as are water, carbon dioxide and most organic substances. Many metals are also diamagnetic, and this requires a special explanation. Take, for example, the simple metal, sodium.

When sodium forms a salt, for example NaCl, it gives up its outermost electron to the chlorine atom, so that both the resulting ions, Na^+ and Cl^-, have noble gas type electron configurations. As expected, NaCl is diamagnetic. But in sodium metal only ten electrons in each atom can be 'paired off'. The eleventh electron from each atom is free to wander about in the metal, as shown by the fact that sodium conducts electricity. We should expect these electrons, with their spin magnetic moments, to cause sodium to be paramagnetic. But it turns out that metals are a law unto themselves. The very fact that the electrons move about in the metal, enables them to pair off their spins so that almost exactly half of them are parallel and the rest antiparallel to any applied field. It is like an extension of the principle causing the H_2 molecule to have no magnetic moment, to the 'giant molecule' of a piece of sodium. Although there is a small amount of paramagnetism resulting from the spins of the free electrons, it is not sufficient to counteract the diamagnetism of the ionic cores in sodium. Many other metals, for example copper, mercury and bismuth, are diamagnetic for similar reasons.

Permanent magnetic moments are found mostly in atoms and ions of metals in three groups of the periodic table: (1) transition metals, e.g. chromium, manganese, iron, cobalt, nickel and copper, (2) rare earth metals, and (3) actinides. These elements and their compounds make up most of the materials that are paramagnetic, ferromagnetic, antiferromagnetic and ferrimagnetic. Many of the salts of transition metals and rare earth metals, for example, are paramagnetic. Many more that belong to one of the other classes when in the crystalline state, exhibit paramagnetism in solution. Most rare earth oxides are also paramagnetic.

Ferromagnetism occurs in nine elements: three transition metals, iron, cobalt and nickel, with atomic numbers $Z = 26$, 27 and 28 respectively, and six rare earth metals, gadolinium, terbium, dysprosium, holmium, erbium and thulium, with Z from 64 to 69. Most alloys consisting of the three transition metals are ferromagnetic, and so are many of their alloys with non-magnetic elements. Many alloys of manganese ($Z = 25$) with non-magnetic elements are also ferromagnetic, although manganese is not ferromagnetic in the pure state.

Antiferromagnetism occurs most commonly in transition metal oxides, such as MnO, CoO and NiO, as well as in other similar compounds such as sulphides and selenides, and some other compounds, for example $CuCl_2$.

Ferrimagnetism occurs mainly in mixed oxides of iron and of other elements, and also in two of the oxides of iron, Fe_3O_4 (already mentioned as magnetite or lodestone) and γ-Fe_2O_3 (called *maghemite*).

The rest of this chapter describes the properties of materials of each main class in more detail.

2.2 Paramagnetism

From the point of view of practical applications, diamagnetic materials are not of interest. The quantitative explanation of their behaviour in magnetic fields, given in the last chapter, is sufficient and we need not go into further details. Paramagnetic materials are similarly of little interest to us for their own sake. However, some understanding of their properties is necessary because it will help us in discussing ferromagnetism. (Paramagnetic materials have one important application, in the production of very low temperatures. However, the scope of this book does not extend to a discussion of this topic.)

2.2.1 The magnetization of paramagnetic materials

It has already been mentioned (section 2.1.3) that paramagnetic materials contain atoms with permanent magnetic dipole moments. In zero applied magnetic field, the moments point in random directions. When a field, H, is applied, a small magnetization, M, develops, but the susceptibility, χ, is very small, and is inversely proportional to the absolute temperature. These are the most important properties of paramagnetic materials. We will now give a brief explanation of these properties.

First of all, if we place an isolated atom with a magnetic moment in a magnetic field, it turns out that its magnetic moment does not line up in the direction of the field. Instead, it rotates about the direction of the field, so that the angle between the field and the moment remains constant. This rotation is called *precession*, and is similar to the behaviour of a spinning top. The top, being acted on by a gravitational field, precesses about the direction of the field, i.e. the vertical. However, eventually the precession ceases, because of friction and air resistance, and the top falls over. In a similar way, if we could slow down the precession of the atomic moment, it would be able to align itself with the field.

As soon as the atom is placed among other, similar atoms, it will interact with them, and it will not be able to continue to precess indefinitely. In an assembly of atoms, such as a gas, liquid or solid, there are enough interactions between the atoms for the precession to stop in a very short time after the application of a magnetic field. In a measurement of M, we usually cannot tell that any precession has taken place. Why is it then that the susceptibility of paramagnetic materials is so small? When a magnetic field is applied, why do the moments not all become parallel to the field? If they did, the magnetization would be Nm, where N is the number of magnetic dipoles, of moment m, per unit volume. But the magnetization, M, is generally several orders of magnitude smaller than Nm. Thus, the direction of the magnetic moments is still almost completely random even when a field is present. All that the field can achieve is a very slight excess in the number of moments pointing parallel to it, over the number pointing in the opposite direction.

The clue to the explanation is the temperature dependence of χ. The higher the temperature, the more difficult it becomes to align the moments. The alignment must therefore be opposed by thermal agitation. Combining together the effect of thermal agitation and the applied magnetic field, we can deduce the behaviour of the assembly of magnetic moments from simple arguments based on thermodynamics. We will not discuss the derivation here, but merely give the result. It predicts that the magnetization of all paramagnetic materials follows a curve similar to the one illustrated in Fig. 2.3. In this figure, we have plotted the dimensionless parameter, M/Nm, as a function of another dimensionless parameter, $\mu_0 mH/kT$, where k is a universal constant taken from thermodynamics, called *Boltzmann's constant*, which has the value $\approx 1.38 \times 10^{-23}\,\mathrm{J\,K^{-1}}$ (see appendix A). For convenience, we will denote the parameter $\mu_0 mH/kT$ by the symbol α. We shall see that for small values of α, the magnetization is proportional to α, whereas for large values of α, M/Nm tends to the constant value 1 for all materials. The behaviour at large α is as expected, as it corresponds to complete alignment of the magnetic dipoles in the direction of the field.

We can easily estimate the value of α under normal laboratory conditions. An average size electromagnet could produce a field of, say, $1\,\mathrm{T}$, so if we take $m = 1\mu_B$ and $T = 300\,\mathrm{K}$, we get $\alpha = 2.2 \times 10^{-3}$. Thus, M will be a very small fraction of the saturation value, Nm. For such small values of α, M is proportional to α, i.e. to H/T. We can write

$$\frac{M}{Nm} \approx c\frac{\mu_0 mH}{kT}, \tag{2.3}$$

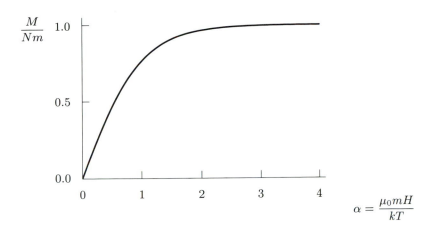

FIG. 2.3   Variation of the magnetization of a typical paramagnetic material with applied field and temperature.

i.e.

$$M \approx \frac{c\mu_0 N m^2 H}{kT}, \tag{2.4}$$

where c is a constant. Its value can be shown to be always between $\frac{1}{3}$ and 1. (The relationship between the formulae in this chapter and those used in other texts is given in appendix B.) If we compare eq. (2.4) with eq. (1.9), we see that the susceptibility of paramagnetic materials is given by

$$\chi = \frac{c\mu_0 N m^2}{kT}. \tag{2.5}$$

Eq. (2.5) can now be compared with eq. (2.2), showing that for paramagnetic materials, eq. (2.2) applies with $\theta = 0$ and

$$C = \frac{c\mu_0 N m^2}{k}. \tag{2.6}$$

Eq. (2.5) is called the *Curie law*, and C is the *Curie constant*. For an ideal paramagnetic material, the Curie law is obeyed at all temperatures, except that at very low temperatures it is only obeyed if the applied field is not too large. At very low temperatures, around the boiling point of helium (4.2 K) and in very large fields, large values of α can be reached and saturation effects may be observed.

Measuring the properties of paramagnetic materials enables the atomic magnetic moment, m, to be estimated. This can be done in two different ways. First, we can measure the susceptibility at a known temperature, and then obtain m from eq. (2.5). For this, we must however know the value of c. A more direct method is to measure the saturation magnetization, which

21

is given by Nm. Experimentally, this is more difficult, since it requires very low temperatures and large fields.

Eqs (2.3) and (2.4) describe the behaviour of paramagnetic materials shown in Fig 2.3, for small values of α. The entire curve of Fig. 2.3 may be represented as

$$M = Nmf_c(\alpha), \qquad (2.7)$$

where f is function of α whose exact form depends on the value of c. For small α, $f_c(\alpha)$ reduces to $c\alpha$, and for large α, $f_c(\alpha)$ tends to 1 for all c. Apart from these two extreme cases, we need not go into details about the exact shape of $f_c(\alpha)$. (For readers who wish to know these details, they are given in appendix B.) However, the fact that $f_c(\alpha)$ tends to 1 when α is large is important because it enables us to measure m, the magnetic moment of the atoms by applying very large fields at low temperatures. Thus, as part of the discussion of paramagnetism, we need to go into rather more detail about the origin of the magnetic moments of atoms. This will not only tell us something about the value of the constant c for various materials, but will also be important in the discussion of the other classes of materials.

2.2.2 Electron orbits, spin and magnetism

In section 2.1.3 it was explained that each electron contributes to the magnetic moment of the atom because of its orbital motion and its spin. The example of the helium atom was used to show that the magnetic moment of closed electron shells is zero. In order to have a net magnetic moment, the atom must have an unfilled electron shell. To illustrate the way the magnetic moment can be predicted, we will use the series of elements with atomic numbers from 21 (scandium) to 30 (zinc). This series contains the most important elements from the point of view of magnetism. In all these elements, 20 electrons in each atom are in closed shells: 2 electrons are in the $1s$-shell, 2 in the $2s$-shell, 6 in the $2p$-shell, 2 in the $3s$-shell, 6 in the $3p$-shell and 2 in the $4s$-shell. (The nomenclature for the electron shells is taken from atomic physics and spectroscopy.) The remaining electrons go into the $3d$-shell. This shell can accommodate up to 10 electrons. In zinc, all 10 electrons are present, the $3d$-shell is closed, and the total magnetic moment is zero. In the other nine elements, the $3d$-shell is incomplete, and we therefore expect these elements to have a magnetic moment.

We can give some simple, empirical rules for working out the magnetic moment. For detailed proofs, we would have to go into the quantum theory of atoms in some detail, but there is no need for us to do that. We have already seen that the electron spin gives rise to a magnetic moment of $1\mu_B$. For an electron in the $3d$-shell, the orbital motion contributes a magnetic moment of $2\mu_B$. What we need to know is the relative orientation of the magnetic moment vectors of different electrons, so that we may add the moments vectorially. This is what we need our empirical rules for. The first rule says that the magnetic moment vectors can only point in certain specified directions. The orbital magnetic moments must orient themselves so that their component in a fixed direction, as defined, for example, by an applied magnetic field, is a multiple of $1\mu_B$. There are therefore five allowed values, $-2\mu_B$, $-1\mu_B$, 0, $+1\mu_B$, and $+2\mu_B$. The spin magnetic moment can only have two possible orientations, with components $+1\mu_B$ or $-1\mu_B$. There are therefore ten possible combinations of components in the fixed directions. Quantum mechanics tells us that only one $3d$ electron in each atom can have each possible combination of components. (This is a consequence of the *Pauli exclusion principle*.) This confirms immediately that in zinc, the moments of the ten $3d$ electrons just cancel out. However, when there are fewer than ten $3d$ electrons, we have to decide which of the possible components occur and which do not. We use another set of rules for this, called *Hund's rules*. The first rule is that the total spin magnetic moment should be as large as possible. Thus, when there are two $3d$ electrons (titanium, for which $Z = 22$), the spins are parallel, giving a total spin moment of $2\mu_B$. With three $3d$ electrons (vanadium, with $Z = 23$), the spin moment is $3\mu_B$, with four $3d$ electrons (chromium, with $Z = 24$), it is $4\mu_B$, and with five $3d$ electrons (manganese, with $Z = 25$), it is $5\mu_B$. Now, the sixth and subsequent $3d$ electrons cannot have spin moments adding to the moments of the first five, since there are only five different orbital states. So, the sixth $3d$ electron must contribute a spin moment of $-1\mu_B$, making the total $4\mu_B$ (for iron, with $Z = 26$). Now, each subsequent $3d$ electron contributes a spin moment of $-1\mu_B$, so the total spin moment decreases with each further increase of atomic number; it is $3\mu_B$ for cobalt ($Z = 27$), $2\mu_B$ for nickel ($Z = 28$), $1\mu_B$ for copper ($Z = 29$) and 0 for zinc ($Z = 30$).

Hund's second rule is that the total magnetic moment due to orbital motion should also be a maximum, subject to having already satisfied the first rule. Thus, the orbital contribution to the moment in scandium is $2\mu_B$. In titanium, we cannot have another orbital contribution of $2\mu_B$, since this would mean that two electrons have the same orbital as well as spin magnetic moments. To satisfy the first as well as the second of Hund's rules, the second electron must contribute $1\mu_B$ to the orbital magnetic moment, making a total of $3\mu_B$. When there are three $3d$ electrons, the third contributes 0, making the total again $3\mu_B$; with four $3d$ electrons, the fourth contributes $-1\mu_B$, making the total $2\mu_B$, and with five $3d$ electrons, the fifth electron, with $-2\mu_B$, brings the total back to 0. From six to ten $3d$ electrons, the sequence repeats itself, so that the total orbital magnetic moment is $2\mu_B$ for iron, $3\mu_B$ for cobalt and the same for nickel, $2\mu_B$ for copper and 0 for zinc.

We can also work out the magnetic moments of ions of the above elements. In doubly charged (divalent) ions, it is the two $4s$ electrons that are given up, so the magnetic moment is the same as that of the uncharged atom. In triply charged (trivalent) ions, one of the $3d$ electrons is also given up, so that the magnetic moment is the same as that of an uncharged atom of the previous element of the periodic table. For example, the magnetic moment of Fe^{3+} is the same as that of Mn (and of Mn^{2+}).

Having given the rules for working out the total magnetic moment arising from spin and from orbital motion respectively, we would need to know how to combine them together to find the total magnetic moment of the atom. In free atoms and ions, according to Hund's third rule, the two contributions are subtracted if the electron shell is less than half full, and added if it is more than half full. This rule applies to atoms in vapours, and to atoms and ions in compounds, provided the compound is molten or in solution. In solids, it is found by experiment that the measured magnetic moment is very close to that given by the spin contribution alone. This phenomenon is called *quenching* of the orbital moment. Very briefly, it is a consequence of the fact that the electron orbitals are fixed in the crystal lattice, and cannot change their orientation when a magnetic field is applied. In the cases we shall be concerned with, it will be sufficient to regard the magnetic moment as being due to the spin only. It is usual in these cases to use the terms *spin* and *magnetic moment* as if they were synonymous.

A second important group of magnetic elements are the rare earths. In these elements, the magnetic moment is due to the $4f$ electrons. The $4f$ shell can accommodate up to 14 electrons, so we can calculate the magnetic moments from Hund's rules. Up to seven electrons can have parallel spins, and the orbital contribution of the electrons can take values from $+3\mu_B$ to $-3\mu_B$. In rare earth atoms and ions, the orbital moment is not quenched even in crystalline solids. It is thought that this is because the $4f$ electrons are more effectively shielded by outer electrons than the $3d$ electrons are in transition metals.

2.3 Ferromagnetism

2.3.1 General description

Ferromagnetic materials differ from diamagnetic and paramagnetic materials in very striking ways. We need sensitive measuring apparatus to detect diamagnetism and paramagnetism in a material. Ferromagnetism can however be very easily detected, which explains why the phenomenon has been known for so many centuries. As stated in section 2.1.2, ferromagnetic materials acquire a large magnetization in relatively small magnetic fields. This magnetization corresponds to all atomic magnetic moments being aligned. In section 2.2.1, it was shown that in a magnetic field of the order of 1 T, at room temperature, the magnetization of paramagnetic

materials was only a small fraction of the saturation value. Most ferromagnetic materials would be saturated in these conditions; many of them
would be saturated in fields very much less than 1 T. (We must make an
exception in the case of some materials in which ferromagnetic behaviour
occurs only below room temperature.) The value of the saturation magnetization varies with temperature, T, decreasing from a maximum value
at $T = 0$ K at first slowly, then more and more rapidly as T increases,
becoming zero at the Curie temperature, θ_C. Above θ_C, the behaviour is
similar to that of a paramagnetic material, with the magnetization being
proportional to the field and the susceptibility decreasing with increasing
temperature. However, the susceptibility does not follow the Curie law,
eq. (2.5), but varies as

$$\chi = \frac{C}{T - \theta} \tag{2.8}$$

(cf. eq. (2.2)), where C is a constant and θ is approximately equal to θ_C.
Eq. (2.8) is called the *Curie-Weiss law*.

The behaviour so far outlined suggests that in ferromagnetic materials,
the atomic magnetic moments have a strong tendency to be aligned parallel
to each other. This tendency aids the external field in producing saturation,
i.e. complete alignment, and also increases the susceptibility by introducing
the term $-\theta$ in eq. (2.8). However, one more well-known experimental result
has to be explained: ferromagnetic materials do not remain saturated when
the applied field is removed. The magnetization in zero field can have
a range of values including zero. Thus, in addition to the tendency for
the magnetic moments to be aligned, there is another effect that tends to
counteract the alignment.

2.3.2   The Weiss theory

The first theory to explain all aspects of this behaviour was developed by
Weiss at the beginning of this century. He introduced the concept of a
molecular field, present in ferromagnetic materials. He assumed that the
molecular field was very large, its magnitude was independent of any exter-
nally applied field, and its direction was not fixed, but was always parallel
to the magnetization. If the magnetization direction rotates under the action of an applied field, the direction of the molecular field rotates with it.
The Weiss theory assumes that the molecular field, H_m, is proportional to
the magnetization:

$$H_m = wM, \tag{2.9}$$

where w is a constant, called the *Weiss constant*. It has a very large value
so that below the Curie temperature, a very large molecular field is present.
Above the Curie temperature, M is zero when there is no field applied, so
that H_m is also zero. As soon as a field is applied, M acquires a small non-
zero value, so that a non-zero molecular field appears to aid the applied

field in increasing the magnetization. This is the origin of the $-\theta$ term in eq. (2.8).

In order to explain the fact that ferromagnetic materials do not remain saturated when the applied field is removed, Weiss introduced the concept of *magnetic domains.* He postulated that a ferromagnet is divided into regions (domains), within which the magnetization is equal to the saturation value. The magnetization in different domains is in different directions, so that the magnetization of a ferromagnetic specimen could be small or even zero. Saturation is produced by aligning the magnetization of each domain with the applied field.

The molecular field concept and the domain hypothesis are the two important ideas on which an understanding of ferromagnetism is based. We must therefore discuss them in more detail.

In section 2.2, it was shown that the magnetization of a paramagnetic material could be expressed as a function of the dimensionless parameter α, equal to $\mu_0 mH/kT$. The dependence of M on α could be expressed in functional form, as in eq. (2.7), or illustrated graphically, as in Fig. 2.3. We can make these results fit ferromagnetic materials by introducing an apparently small modification. In the definition of α, the magnetic field H appears. In the case of ferromagnetic materials, H must include the molecular field as well as the applied field. Hence we must write

$$\alpha = \frac{\mu_0 m(H + wM)}{kT}.\tag{2.10}$$

Then eq. (2.7) becomes

$$M = Nmf_c[\mu_0 m(H + wM)/kT].\tag{2.11}$$

In this equation, f_c means a function whose value depends on the value of the argument, $\mu_0 m(H+wM)/kT$. As Fig. 2.3 shows, f_c approaches 1 when the argument is large, and is approximately equal to c times the argument when the argument is small. We wish to find the value of M for given values of H and T. The solution can in principle be found from eq. (2.11). However, it turns out that the function f_c is too complicated for eq. (2.11) to be solved analytically. A numerical solution is always possible, and could in fact be obtained fairly quickly even on a programmable pocket calculator. However, we are more interested in predicting the general behaviour than in obtaining numerical solutions for particular cases. We can get a very good idea of the general behaviour by using a simple graphical method. The method makes use of the fact that both eqs (2.10) and (2.11) can be regarded as expressing the dependence of M on α. To make this point clearer, we solve eq. (2.10) for M:

$$M = \frac{kT\alpha}{\mu_0 mw} - \frac{H}{w},\tag{2.12}$$

and write eq. (2.11) in the form

$$M = Nmf_c(\alpha). \tag{2.13}$$

To find a solution for M, we need to find the value of α which gives the *same* M from eqs (2.12) and (2.13). In other words, if we plot a graph of M against α from both equations, the solution will be at the point of intersection of the two graphs (or points of intersection if there is more than one intersection). The graphs turn out to be quite simple. Eq. (2.13), which is identical to eq. (2.7), has already been plotted (Fig. 2.3). The precise form of the curve depends on c, but c is constant for a given material. The important point is that the same curve is valid at all temperatures and applied fields, since T and H do not appear explicitly in eq. (2.13).

Plotting eq. (2.12) is even simpler. The graph is always a straight line. Its slope is proportional to T, and its intercept to H. Thus, the graphical method consists of plotting a single curve to represent eq. (2.13), superimposing on it various straight lines corresponding to different values of T and H, and finding the intersection in each case. Let us examine some specific cases.

When there is no applied field, eq. (2.12) reduces to

$$M = \frac{kT\alpha}{\mu_0 mw}, \tag{2.14}$$

so that the straight line always passes through the origin. Fig. 2.4 shows the behaviour at different temperatures. Curve A represents eq. (2.13). The straight lines B_1 to B_5 represent eq. (2.14). Line B_1 corresponds to a low temperature. The intersection with curve A occurs at a large value of α, where M is very nearly equal to Nm. At a higher temperature, we reach line B_2. The value of α corresponding to the intersection is considerably smaller, but we are still on a part of curve A where M is nearly constant, so that M has not decreased significantly from its maximum value of Nm. As the temperature increases further, we reach line B_3. Now M has started to decrease, and it obviously decreases more and more rapidly with increasing temperature. Eventually, M becomes zero at the temperature at which the straight line is a tangent to curve A. This is line B_4, and it corresponds to the Curie temperature, θ_C. We can estimate θ_C, because curve A can be approximated by $c\alpha$ for small values of α. Hence eq. (2.13) becomes

$$M = Nmc\alpha. \tag{2.15}$$

Combining eqs (2.14) and (2.15) gives

$$T = \frac{\mu_0 Nm^2 cw}{k} = \theta_C. \tag{2.16}$$

It is seen that the molecular field coefficient w is proportional to θ_C.

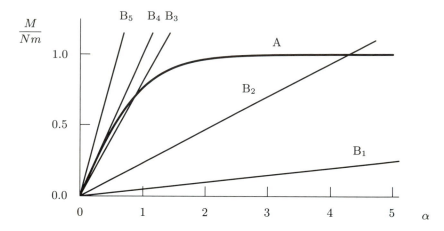

FIG. 2.4   Graphical determination of magnetization for a ferromagnet in zero
applied field. Curve A is a plot of eq. (2.13), and lines B_1, B_2, B_3, B_4
and B_5 are plots of eq. (2.12) at different temperatures.

Above θ_C, the only solution is $M = 0$, as in the case of the line B_5. It
should be noted that below θ_C, there are two intersections, one at $M = 0$
and one at $M > 0$. Both solutions for M represent states of equilibrium.
However, it can be shown that the solution at $M = 0$ corresponds to unstable equilibrium, while that at $M > 0$ corresponds to stable equilibrium.
Thus, at temperatures below θ_C, ferromagnetic materials always have a
non-zero, spontaneous magnetization.

The variation of spontaneous magnetization with temperature for ferromagnetic materials has already been illustrated (Fig. 2.1). We have now
seen how this variation is derived from the graphical method of Fig. 2.4. It
has also been mentioned that Fig. 2.1, which relates the two dimensionless
quantities $M_s(T)/M_s(0)$ and T/θ_C, is followed approximately by all ferromagnetic materials. In fact, this is only approximately true; the shape
of the curve varies slightly according to the value of c. However, most
ferromagnetic materials follow the curve appropriate to $c = 1$ fairly closely.

We can now go on to discuss what happens when an external field is
applied. As eq. (2.12) shows, the effect is to displace the line B to the right
by a distance proportional to H. The effect of an applied field at three
different temperatures is illustrated in Fig. 2.5. Line B_2 corresponds to zero
field at a lower temperature. When a field is applied, the line is displaced to
B_2'. The value of α corresponding to the intersection with curve A increases
but the magnetization does not change significantly, since it is already near
its maximum value of Nm even in the absence of the applied field.

At a higher temperature, the behaviour in zero field is described by
line B_3. But now the intersection occurs on a part of curve A where M is

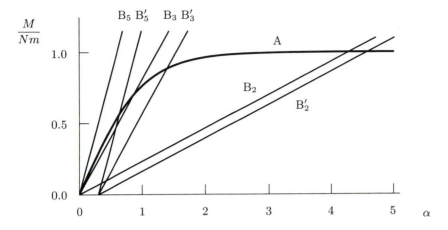

FIG. 2.5   Graphical determination of magnetization for a ferromagnet in an applied field.

not constant. Hence, the applied field will now produce a more significant increase in M (line B_3'). Line B_5 represents the behaviour for $T > \theta_C$ in zero field: the magnetization is zero. When a field is applied, a non-zero magnetization is produced (line B_5'). We can estimate the magnetization from eqs (2.12) and (2.15). Eliminating α between the equations, we get

$$M = \frac{kTM}{\mu_0 Nm^2 cw} - \frac{H}{w}. \tag{2.17}$$

The susceptibility is given by M/H, so that

$$\chi = \frac{\mu_0 Nm^2 c}{kT - \mu_0 Nm^2 cw}, \tag{2.18}$$

which can be written as

$$\chi = \frac{\mu_0 Nm^2 c/k}{T - \mu_0 Nm^2 cw/k}. \tag{2.19}$$

Eq. (2.19) is of the same form as eq. (2.8), with

$$C = \frac{\mu_0 Nm^2 c}{k} \tag{2.20}$$

and

$$\theta = \frac{\mu_0 Nm^2 cw}{k}. \tag{2.21}$$

This shows that the Curie-Weiss law, eq. (2.8), is one of the results derived from the Weiss theory of ferromagnetism. We also see that θ is in fact equal to the Curie temperature (cf. eqs (2.16) and (2.21)).

An important point needs to be emphasized concerning the magnitude of the fields we have been considering. As we have said before (section 2.3.1), the magnetization of ferromagnetic materials is smaller, usually very much smaller than the saturation value, in zero field, as a consequence of the subdivision of the materials into domains. In the discussion of the temperature dependence of the saturation magnetization in zero field, it was assumed that we were referring to the magnetization of individual domains, not to the average magnetization of the specimen. Alternatively, we could have assumed that a small field was present, sufficient to align the domain magnetizations, but not large enough to affect the saturation magnetization. When we extended the discussion to the effect of applied fields, we were referring to much larger fields than the field necessary to saturate the specimen at temperatures below the Curie temperature.

For a more complete understanding of ferromagnetic materials, we shall have to discuss the domain theory in more detail. In fact, Weiss did not suggest a reason for the subdivision of ferromagnetic materials into domains. He merely stated that the existence of domains could explain the large changes of magnetization caused by relatively small fields. It was only about 30 years after the publication of the Weiss theory that the reasons for the existence of domains were explained. We will leave the discussion of domain theory until later (section 3.2).

2.3.3 Comparison with experimental results

To complete the discussion of the Weiss theory, we need to examine how closely it agrees with experimental results. It is found that in most ferromagnetic materials, the variation of the spontaneous magnetization with temperature follows the predicted behaviour, Fig. 2.1, fairly well, except near the Curie temperature. Similarly, the Curie-Weiss law, eqs (2.19), (2.20) and (2.21), holds fairly well at high temperatures, but slight deviations from eq. (2.19) occur near the Curie temperature. In fact, if we calculate θ from measurements of χ at high temperatures, it will be found that the value is not quite the same as the temperature at which the spontaneous magnetization disappears. These two temperatures are usually referred to as the *paramagnetic* and the *ferromagnetic* Curie temperature respectively. As an example, in gadolinium, the two Curie temperatures have been measured as $(317 \pm 3)\,\mathrm{K}$ (paramagnetic) and $(293.2 \pm 0.4)\,\mathrm{K}$ (ferromagnetic). The slight discrepancy between the Weiss theory and experimental results is due to the assumption implied by eq. (2.9). In this equation, M is assumed to be the average magnetization of the specimen. We clearly cannot take this assumption too literally: in a demagnetized specimen, the magnetization averaged over the whole volume is zero, yet we know that a large molecular field is present. We must therefore change our assumption slightly, and say that M is the average magnetization within a given domain. But it turns out that even this assumption is incorrect. Although our understanding of the origin of the molecular field is still very incomplete, it is now generally accepted that the forces producing the alignment of magnetic moments act over very short distances only—distances of the order of the interatomic distance. So, H_m at a particular point in the specimen is proportional to the magnetization in the immediate neighbourhood of that point, and not to the magnetization averaged over many atoms. This explains why the Weiss theory is much more satisfactory at very low and at very high temperatures than near the Curie temperature. At low temperatures, there is a high degree of alignment of the magnetic moments within the domains, and there is no significant difference between *average* and *local* magnetization. At high temperatures, the magnetic moments are almost completely disordered even on a local scale. Near the Curie temperature, the magnetic moments of small groups of atoms are still fairly well aligned, but the direction of the alignment changes over fairly short distances. There is therefore still quite a large local magnetization while the average magnetization is quite small. Unfortunately, it is very much more difficult to derive a theory of ferromagnetism based on a molecular field related to local magnetization than the theory we have discussed so far. But the Weiss theory provides a simple explanation for many aspects of ferromagnetism.

There is one further difficulty concerning the theory of ferromagnetic materials. In section 2.1.3 we discussed the way in which the magnetic moments of atoms could be calculated from a knowledge of their electronic structure. The magnetic moment per atom can be determined experimentally by measuring the saturation magnetization and dividing it by the number of atoms per unit volume. For the ferromagnetic transition metals, the resulting values are $2.2\mu_B$ per atom in iron, $1.8\mu_B$ per atom in cobalt and $0.6\mu_B$ per atom in nickel. These are much smaller than the values estimated from the electronic structure of the atoms or ions, and moreover, they are far from being integral numbers of Bohr magnetons. A simple explanation of this paradox is that in these metals, the electrons that are responsible for the magnetic moments are not bound to particular atoms, but are fairly free to move about in the metal. For such cases, the magnetic moment per atom is governed by rather different principles from those described in section 2.1.3. However, the magnetic moments of rare earth ferromagnets are much closer to the values we may expect for free atoms. In these materials, the electrons carrying the magnetic moments are fairly well localised at the atoms.

2.4   Antiferromagnetism and ferrimagnetism

We have already seen (section 2.1.2) that ferrimagnetic materials have properties that are in many ways similar to those of ferromagnetic materials,

31

making them useful for many practical applications. However, there are also important differences between the properties of the two classes. The properties of ferrimagnetic materials cannot be explained by the Weiss theory outlined in section 2.3.2. In order to discuss the modifications that need to be made to the Weiss theory to make it applicable to ferrimagnetic materials, it is useful to discuss antiferromagnetic materials first. Antiferromagnetic materials do not have technologically useful magnetic properties, but they can be regarded as being special, simple types of ferrimagnetic materials. The theory of both antiferromagnetism and ferrimagnetism was first developed by Néel.

2.4.1 Properties of antiferromagnetic materials

Antiferromagnetic materials were originally thought of as a class of anomalous paramagnets, since they have small positive susceptibilities of similar magnitude to many materials of the latter class. As explained in section 2.1.2, the anomaly manifests itself in the variation of the susceptibility with temperature, eq. (2.2) with θ having a positive sign being obeyed above a critical temperature, called the *Néel temperature*, and a more complicated variation below it.

The clue to the explanation of this behaviour was provided by neutron diffraction experiments. Neutrons have a magnetic dipole moment, and when they are diffracted from a crystalline material, the resulting pattern contains information not only about the periodic arrangement of the atoms (as an X-ray diffraction pattern does), but also about any periodicity in the arrangement of the magnetic moments. The information provided by neutron diffraction can be illustrated by considering a typical simple antiferromagnet, MnO. The crystal structure is NaCl type. A unit cell is illustrated in Fig. 2.6. Above the Néel temperature, the neutron diffraction pattern corresponds to the periodic arrangement of Mn^{++} and O^{--} ions. However, when the specimen is cooled below the Néel temperature (122 K), extra reflections appear in the pattern. These reflections can be interpreted as arising from the fact that the neutrons now 'see' two different types of Mn^{++} ions, whose arrangement on the lattice is also periodic. The orientations of the magnetic moments of the Mn^{++} ions have become fixed, and the moments of ions of one type point in the opposite direction to those of ions of the other type. The lattice of Mn^{++} ions is now subdivided into two sublattices, termed A and B.

As Fig. 2.6 shows, the moments of nearest neighbour Mn^{++} ion pairs are not necessarily antiparallel. (The nearest neighbour pairs lie in directions parallel to cube face diagonals relative to each other.) In fact, just as many nearest neighbour pairs have parallel moments as have antiparallel moments. However, all next nearest neighbour pairs, which lie along cube edge directions with an O^{--} ion between them, have antiparallel moments. This suggests that the O^{--} ions must play a part in the forces

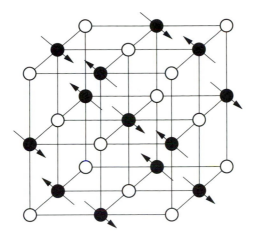

FIG. 2.6 Unit cell of MnO. Open circles are O^{--} ions, full circles are Mn^{++} ions. The arrows show the alignment of magnetic moments below the Néel temperature. Ions belonging to the A and B sublattices have magnetic moments pointing 'up' and 'down' respectively.

producing the antiparallel alignment of magnetic moments. For a detailed understanding of these forces, we would need a complicated, quantum-mechanical theory. Without going into such details, we can get some idea of the way these forces work. The O^{--} ion has six outermost electrons. These electrons are arranged in three orbitals of $2p$-type, with two electrons to each orbital. The spins of the two electrons in the same orbital are antiparallel. Each $2p$-orbital has an elongated shape, rather like an hourglass. The axes of the three orbitals are mutually perpendicular, each aligned in the direction of one of the cube edges in the crystal. Thus, each orbital reaches out towards a pair of nearest neighbour Mn^{++} ions. Each Mn^{++} ion has five $3d$ electrons, with spins all parallel, according to Hund's rules. Now, we should imagine that the $2p$-electrons of the O^{--} ion occasionally hop onto the Mn^{++} ion which happens to be near the end of the $2p$ 'hourglass'. But a sixth electron in the $3d$ orbital of the Mn^{++} must have opposite spin to the original five $3d$-electrons. Thus, the net magnetic moment of the Mn^{++} ion is in the opposite direction to the spin of the electron from the O^{--} ion. At the same time, the other electron in the same $2p$-electron orbital of the O^{--} ion is interacting with the Mn^{++} ion at the opposite end of the 'hourglass'. Since the two $2p$-electrons have opposite spins, the result is that the magnetic moments of the two Mn^{++} ions are antiparallel to each other. In the MnO structure, there is in fact one 'hourglass' joining each pair of next nearest neighbour Mn^{++} ion pairs. Thus, we have at least a qualitative explanation of the antiparallel align-

33

ment of the moments of all next nearest neighbour Mn^{++} ion pairs. At first sight, it may seem surprising that two ions should interact with each other at a comparatively large distance, without being directly in contact. We now see that the intervening O^{--} ion plays an important part in the interaction. This type of interaction, with a non-magnetic ion playing a part in aligning the magnetic moments of two ions on either side of it, is called *superexchange*.

Perfect alignment of magnetic moments is only realised at absolute zero temperature. At non-zero temperatures, the alignment is counteracted by the available thermal energy. We will now see how the main features of the properties of antiferromagnetic materials can be explained by Néel's theory.

2.4.2 Néel's theory of antiferromagnetism

In the case of ferromagnetism, we based our theory on the assumption that a molecular field, given by eq. (2.9), was present. A similar assumption can be made for antiferromagnetic materials. Here, however, we have to distinguish between the molecular fields acting on ions on the A and B sublattices respectively. The molecular field, H_A, acting on an ion on the A sublattice, for example, depends on the magnetization of both the A and B sublattices, M_A and M_B. We will however assume, for simplicity, that H_A depends on M_B only, not on M_A. Thus we may write

$$H_A = -wM_B, \tag{2.22}$$

where w is the (dimensionless) molecular field constant. In an antiferromagnetic material H_A is in the opposite direction to M_B, and the negative sign in eq. (2.22) allows w to be positive. By symmetry, we must have

$$H_B = -wM_A. \tag{2.23}$$

In zero applied field, we obviously have zero net magnetization at all temperatures. However, when a field H is applied, the total field acting on magnetic moments on A and B sites is $H - wM_B$ and $H - wM_A$ respectively.

In the derivation of the expression for the magnetization of a ferromagnet, eq. (2.11), we followed the method used for paramagnets, leading to eq. (2.7). We only needed to redefine α, to make it include the molecular field. This approach is valid for antiferromagnets as well, but we must remember that we now need separate expressions for M_A and M_B. Let N be the total number of magnetic moments of magnitude m; half of these moments are on A sites and the other half on B sites. In analogy with eq. (2.11) we have

$$M_A = \tfrac{1}{2}Nmf_c[\mu_0 m(H - wM_B)/kT], \tag{2.24}$$

and

$$M_B = \tfrac{1}{2}Nmf_c[\mu_0 m(H - wM_A)/kT]. \tag{2.25}$$

We could not solve eq. (2.11) analytically, and eqs (2.24) and (2.25) are even less promising in this respect: in addition to the previous complications, we now have two equations, both of which contain M_A as well as M_B. But in this case, we can deduce all the important properties of antiferromagnets even without resorting to the graphical method used for ferromagnets.

The simplest case to consider is that of large T. The argument of the function f_c is then small, and as we have seen before, f_c can be replaced with c times its argument. Eqs (2.24) and (2.25) then become

$$M_A = \frac{c\mu_0 Nm^2(H - wM_B)}{2kT}, \tag{2.26}$$

and

$$M_B = \frac{c\mu_0 Nm^2(H - wM_A)}{2kT}. \tag{2.27}$$

These equations are linear and they can easily be solved for M_A and M_B. The solution is

$$M_A = M_B = \frac{CH}{2(T + \theta)}, \tag{2.28}$$

where

$$C = \frac{c\mu_0 Nm^2}{k}, \tag{2.29}$$

and

$$\theta = \frac{1}{2}Cw. \tag{2.30}$$

The net magnetization is $M_A + M_B$, which gives a susceptibility

$$\chi = \frac{M_A + M_B}{H} = \frac{C}{T + \theta}. \tag{2.31}$$

This result is again of the form of eq. (2.2), but now the equation must be taken with the positive sign; eqs (2.22), (2.23), (2.29) and (2.30) show that θ as defined here is always positive.

Another important result can be deduced from eqs (2.26) and (2.27). In zero applied field, the equations can be written, using eqs (2.29) and (2.30), as

$$M_A + \frac{\theta M_B}{T} = 0, \tag{2.32}$$

and

$$M_B + \frac{\theta M_A}{T} = 0. \tag{2.33}$$

We can have a non-zero solution for M_A and M_B if

$$\frac{\theta}{T} = 1. \tag{2.34}$$

35

Eqs (2.32) and (2.33) are then satisfied if

$$M_A = -M_B, \tag{2.35}$$

i.e. if the two sublattices are magnetized in opposite directions. The temperature at which we have non-zero M_A and M_B in zero applied field is the Néel temperature, θ_N, mentioned in section 2.1.2. We see that

$$\theta_N = \theta \tag{2.36}$$

from eq. (2.34). Below θ_N, eq. (2.31) is not obeyed, and χ varies in a more complicated way. To see what happens, at least qualitatively, we consider a single crystal at absolute zero temperature. The magnetic moments are perfectly ordered, as illustrated in Fig. 2.6. First, consider a magnetic field being applied parallel to the moments. Half the moments are already oriented in a favourable direction with respect to the field. The other half are very rigidly held in the unfavourable direction, and therefore no reorientation of moments takes place. The susceptibility is therefore zero. Now consider a field being applied perpendicular to the moments. In this case, the moments are able to rotate slightly towards the applied field, as illustrated in Fig. 2.7. The susceptibility is therefore not zero. It can be shown that in this case, the susceptibility at absolute zero is equal to the susceptibility at the Néel temperature. So, it is reasonable to assume that as the temperature decreases from θ_N to 0, the susceptibility gradually decreases to 0 in fields applied parallel to the magnetic moments, but stays roughly constant in fields applied perpendicular to them. For a polycrystalline antiferromagnetic material, the susceptibility below the Néel temperature is an average of the values that would be obtained in a single crystal of the same material with the field applied parallel and perpendicular to the magnetic moments. If the alignment of the grains is perfectly random, the susceptibility is

$$\chi = \tfrac{1}{3}\chi_{\parallel} + \tfrac{2}{3}\chi_{\perp}, \tag{2.37}$$

where $\chi_{\parallel}$ and $\chi_{\perp}$ are the values of the susceptibility with the field applied parallel and perpendicular to the magnetic moments in a single crystal.

We have seen that for all types of materials discussed so far, at sufficiently high temperatures, χ^{-1} varies linearly with T. We can therefore illustrate the behaviour of different types of materials by plotting χ^{-1} against T. Fig. 2.8 shows a schematic illustration of the behaviour of different materials. The behaviour of antiferromagnetic materials below θ_N is also shown.

2.4.3   Properties of ferrimagnetic materials

Just as antiferromagnetic materials were at first regarded as anomalous paramagnets, ferrimagnetic materials were considered to be anomalous ferromagnets. In fact, the oldest known 'magnetic' material, magnetite, is a

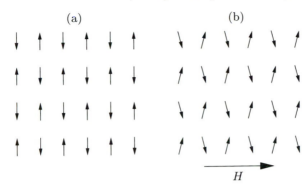

FIG. 2.7   Arrangement of magnetic moments in an antiferromagnet at $T = 0\,\mathrm{K}$, (a) in zero applied field, and (b) in a field, H, applied perpendicular to the moments.

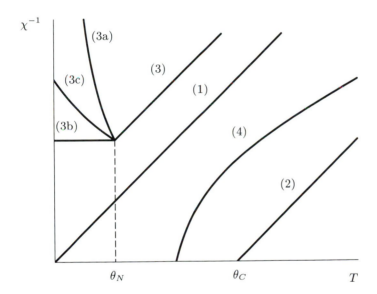

FIG. 2.8   Variation of χ^{-1} with T for (1) paramagnetic, (2) ferromagnetic, (3) antiferromagnetic and (4) ferrimagnetic materials. The behaviour of antiferromagnetic materials in fields applied parallel and perpendicular to the magnetic moments is illustrated as curves (3a) and (3b) respectively. Curve (3c) is an average of (3a) and (3b), appropriate to a polycrystalline antiferromagnet.

ferrimagnet, as already mentioned. Permanent magnetism found in nature is always due to the presence of ferrimagnetic materials, since ferromagnets are generally not stable enough chemically to be preserved for long periods without some artificial protection. Before the nineteenth century,

37

magnetite was the only material known to be capable of having a permanent magnetic moment. With the advent of electrical technology, it became possible to magnetize iron and steel, and when the need for strongly magnetic materials arose for the further development of electrical technology, ferromagnetic materials took over from magnetite in all fields of practical application. During the first half of the twentieth century, ferrimagnetic materials were considered to be useful mainly in helping to develop geological theories. By the 1940s, the main differences between ferromagnetic materials and magnetism occurring in rocks were sufficiently well known for Néel to be able to formulate his theory according to which materials responsible for magnetism in rocks should be considered to belong to an entirely different class from ferromagnets. The name given to the former materials derived from the fact that they always contained iron ions in the trivalent state, i.e. ferric ions. Ferrimagnetic materials began to assume greater importance as electronic technology began to develop. It was found that metallic materials did not have useful magnetic properties in applied fields oscillating at high frequencies. The need for electrically insulating magnetic materials inspired a successful search for artificially produced ferrimagnetic materials. Since the 1950s, a great variety of ferrimagnetic materials have been developed. Some of these materials even have applications other than as cores for coils operating at high frequencies.

The first clue to the explanation of ferrimagnetism is given by the values of the magnetic moments. Unlike ferromagnetic transition metals, most ferrimagnets have magnetic moments that are fairly close to integral numbers of Bohr magnetons per chemical formula units. However, the actual number of Bohr magnetons is always less than the sum of the contributions of the various ions in the formula unit. For example, in magnetite the formula unit is Fe_3O_4. This unit contains one Fe^{++} ion with a moment of $4\mu_B$ and two Fe^{3+} ions with moments of $5\mu_B$ each. Thus we may expect the magnetic moment of the formula unit to be $14\mu_B$. In practice, it is quite close to $4\mu_B$. This value could be explained if the moments of the two Fe^{3+} ions were antiparallel, so that the net moment of the formula unit is given by the moment of the Fe^{++} ion. As in the case of antiferromagnetism, the directions of magnetic moments in ferrimagnetic materials can be determined by neutron diffraction. It has been confirmed in this way that in magnetite, the magnetic moments of half the Fe^{3+} ions are antiparallel to the moments of the other half.

2.4.4   Types of ferrimagnetic materials

2.4.4.1   Simple ferrites

The simplest series of ferrimagnetic materials derives from Fe_3O_4, by the replacement of the FeO component by the oxides of other transition metals in the divalent state. We thus obtain compounds with the formula MFe_2O_4,

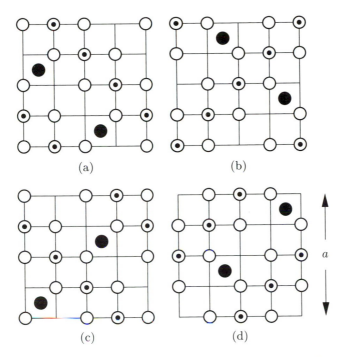

FIG. 2.9   Layer-by-layer illustration of the spinel structure. The diagram shows
the four layers that make up a unit cell of lattice parameter a. Open
circles are O^{--} ions, full circles are A sites and circles with dots are
B sites. A sites are at a height $\frac{1}{8}a$ above each layer. Successive layers
are separated by a vertical distance $\frac{1}{4}a$.

where M is most commonly Mn, Co, Ni, Cu or Zn. In all these materials,
called simple ferrites, the magnetic moment per formula unit is equal to the
magnetic moment of the M^{++} ion (as discussed in section 2.1.3), indicating
that the moments of the Fe^{3+} ions cancel out. (Metallurgists should note
that the word *ferrite* has two quite different meanings: it is used to denote
certain types of mixed oxides containing Fe_2O_3 as one constituent, and also
to denote the body-centred cubic phase in iron and steels. In this book,
the word is always used to denote the former, except in section 3.3.6.)

We can get a better understanding of simple ferrites by considering their
crystal structure. The structure is very similar to that of the mineral spinel,
$MgAl_2O_4$, and we therefore describe the crystal structure as *spinel type*.
The unit cell of this structure is quite large, and it is therefore not easy to
give a three-dimensional representation on a sheet of paper. The easiest
way to describe the unit cell is to take it apart, layer by layer, and draw
the layers side by side, as shown in Fig. 2.9 (in a similar manner to the
three-dimensional version of 'noughts and crosses'!). The unit cell, which is

a cube with edge length a, is therefore built up by stacking each square on top of the one next to it on the left, at distances $\frac{1}{4}a$, $\frac{1}{2}a$ and $\frac{3}{4}a$ above the left-hand square. To complete the cube, we stack a square of the same type as the left-hand one on the top, at a height a. We can now count the number of atoms of each type in the unit cell, remembering that atoms on the face of the unit cell are shared by two cells, atoms on the edges are shared by four cells and atoms on the corners are shared by eight cells. The number of O^{--} ions (marked by crosses in the figure) in the unit cell therefore comes to 32. The O^{--} ions are in a face-centred cubic arrangement. In this arrangement, there are two types of interstitial positions. Consider, first, any of the positions marked by an empty circle. These positions are surrounded on all four sides, as well as above and below, by O^{--} ions. They therefore have six nearest neighbour O^{--} ions, which are arranged on the corners of an octahedron. These interstitial positions are called octahedral sites. Some of these sites are occupied by metal ions. In spinel, these are the Al^{3+} ions, of which there are 16 in the unit cell. Secondly, consider one of the sites marked by a full circle. These sites are not actually in the plane of the paper, but are raised by a distance $\frac{1}{8}a$, so that the sites are half-way between the plane of the square on which they are drawn and the plane of the next square up. These sites are tetrahedral sites, since they are surrounded by four O^{--} ions arranged on the corners of a tetrahedron. In spinel, the tetrahedral sites are occupied by Mg^{++} ions, of which there are 8 in the unit cell.

Any compound in which the corresponding ions are arranged as in $MgAl_2O_4$ is referred to as *normal spinel* type. The crystal structure of the simple ferrites is slightly different. The tetrahedral sites are occupied not by the M^{++} ions but by Fe^{3+} ions. This accounts for half the available Fe^{3+} ions. The remainder, together with all the M^{++} ions, are on the octahedral sites. This structure is called *inverse spinel*. In ferrites, we usually refer to the tetrahedral sites as A sites, and the octahedral sites as B sites.

As in the antiferromagnetic MnO, the metal ions in ferrites are separated by O^{--} ions. It is therefore unlikely that pairs of metal ions can interact directly to produce alignment of their magnetic moments. However, as in MnO, the O^{--} ions play an important part in linking together pairs of metal ions. If that same *superexchange* interaction were at work in ferrites as in MnO, then the tendency for any pair of metal ions linked by an O^{--} ion would be for their magnetic moments to be antiparallel. The strength of the superexchange interaction varies strongly with the angle between the two lines drawn from the O^{--} towards the two metal ions. The interaction increases in strength as this angle increases towards 180°. Unlike in MnO, in ferrites there is no O^{--} ion exactly on a straight line connecting a pair of metal ions. The largest angle is that subtended by an A-site ion and a B-site ion at the O^{--}. This angle is about 125°. The angle subtended by two B-site ions is only 90° and that subtended by two A-site ions is about 80°. The magnetic structure is therefore determined by the requirement that the moments of A sites and B sites must be antiparallel. This is why in the inverse spinel structure the moments of half the Fe^{3+} ions cancel out the moments of the other half, and the net moment is that of the M^{++} ions only. The magnetic moment of various simple ferrites per MFe_2O_4 'formula unit' is therefore expected to be 5, 4, 3, 2, 1 and $0\mu_B$ for Mn, Fe, Co, Ni, Cu and Zn ferrite respectively, and the measured moments are close to these values.

2.4.4.2 Mixed ferrites

Different simple ferrites can be dissolved in each other in various proportions, and the magnetic moment of the resulting mixed ferrite is expected to be given by a linear combination of the magnetic moments of the various M^{++} ions according to their concentration. Important exceptions to this rule are mixed ferrites containing $ZnFe_2O_4$. As shown in Fig. 2.10, the magnetic moment increases at first as $ZnFe_2O_4$ is added. Now, $ZnFe_2O_4$ actually has a normal spinel structure, and as $ZnFe_2O_4$ is added to other ferrites, the structure contains an increasing proportion of normal spinel. Take, for example, the case of mixed nickel-zinc ferrites. Pure nickel ferrite has a magnetic moment of $2\mu_B$ per formula unit, given by the $5\mu_B$ of one of the Fe^{3+} ions cancelling out the $5\mu_B$ of the other, leaving the net $2\mu_B$ of the Ni^{++}. If the same ferrite were normal spinel, the moment would be $8\mu_B$, given by the two Fe^{3+} ions reinforcing and the moment of the Ni^{++} being subtracted. If $NiFe_2O_4$ had a partly normal and partly inverse spinel structure, the moment would be between $2\mu_B$ and $8\mu_B$. By changing $NiFe_2O_4$ from inverse to partly normal spinel, the addition of $ZnFe_2O_4$ increases the moment above $2\mu_B$. But there is another mechanism that contributes to an increase of moment. If a Zn^{++} ion were substituted for a Ni^{++} ion on a B site, the corresponding formula unit would contribute 0 rather than $2\mu_B$ to the moment. However, if a Ni^{++} ion on an A-site were replaced by a Zn^{++} ion, the moment per formula unit would increase from $8\mu_B$ to $10\mu_B$. Zn^{++} ions do have a preference for A-sites, and so this second mechanism also contributes to an increase of moment. In manganese-zinc ferrite, a transformation into normal spinel does not by itself result in an increase of moment, since the moment of Mn^{++} is $5\mu_B$, the same as that of Fe^{3+}. However, an increase does still result from the replacement of Mn^{++} ions on A-sites by Zn^{++} ions.

As Fig. 2.10 shows, the moment of mixed zinc ferrites passes through a maximum and decreases again as the composition approaches pure zinc ferrite. The reason for this is that as Zn^{++} ions have a full $3d$ shell, they cannot interact with O^{--} ions to produce superexchange, and as the concentration of $ZnFe_2O_4$ increases, the tendency of the moments to align becomes weaker and weaker. The addition of zinc ferrite depresses the Curie temperature; if curves similar to those in Fig. 2.10 were plotted at different temperatures, the maxima would move towards higher zinc

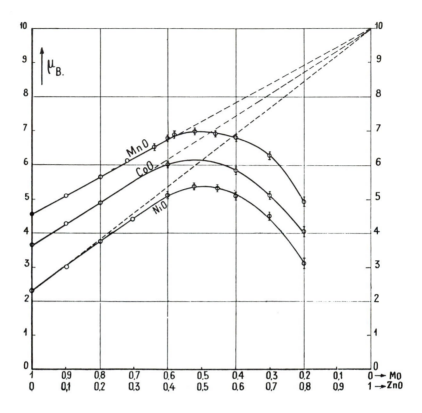

FIG. 2.10   The magnetic moment per formula unit of various ferrites mixed with ZnFe$_2$O$_4$, as a function of composition. (From C. Guillaud, *J. de Phys. et le Radium* **12**, 239 (1951). © Les Éditions de Physique, Les Ulis, France. Reproduced with permission.)

concentration and larger magnetization with decreasing temperature.

2.4.4.3   Hexagonal ferrites

Besides the simple and mixed ferrites described so far, a large variety of ferrimagnetic compounds based on mixtures of Fe$_2$O$_3$ with oxides of divalent metals have been developed. Most of them have crystal structures that are different from, but related to that of spinel. A particularly important class of these ferrites is a range of compounds containing barium oxide, having a hexagonal crystal structure. It is usual to classify them into six types, according to their chemical formulae:

(i) M-type compounds are (MO)(Fe$_2$O$_3$)$_6$,

(ii) W-type compounds are $(BaO)(MO)_2(Fe_2O_3)_8$, which is usually abbreviated to M_2W,

(iii) Y-type compounds are $(BaO)_2(MO)_2(Fe_2O_3)_6$, or M_2Y,

(iv) Z-type compounds are $(BaO)_3(MO)_2(Fe_2O_3)_{12}$, or M_2Z,

(v) X-type compounds are $(BaO)_2(MO)_2(Fe_2O_3)_{14}$, or M_2X,

(vi) U-type compounds are $(BaO)_4(MO)_2(Fe_2O_3)_{18}$, or M_2U.

In these formulae, M stands for a divalent ion which may be non-magnetic such as barium, strontium or lead (particularly in the M-type compounds) or one of the magnetic transition metal ions such as cobalt. Each type of compound has a characteristic crystal structure.

2.4.4.4 Garnets and orthoferrites

Ferrimagnetic materials may also be made from oxides of trivalent ions only. The two most important types are garnets and orthoferrites. Garnets have a general formula $(M_2O_3)_3$ $(Fe_2O_3)_5$, where M is a trivalent metal such as yttrium or a rare earth, or a mixture of more than one of these. They have a cubic crystal structure. A very large number of different garnets have been made and studied in recent years. Some of them contain small amounts of oxides of divalent (e.g. calcium) or tetravalent (e.g. germanium) ions. Orthoferrites have a formula $MFeO_3$, where M is again a trivalent metal such as yttrium or a rare earth. The crystal structure of orthoferrites is similar to that of the mineral perovskite, $CaTiO_3$ (which is also the structure of the well-known piezoelectric material, barium titanate, $BaTiO_3$ used in transducers). However, while perovskite is cubic, orthoferrites are slightly distorted into an orthorhombic structure. They are only weakly magnetic, consisting of two sublattices that are magnetized in almost, but not exactly, antiparallel directions. (This type of structure is called a *canted antiferromagnet*.) In later chapters, we will return to some practical applications of these materials, and we will then describe some of their properties in more detail.

2.4.4.5 Other useful oxides

To complete the list of ferrimagnets, a few special compounds should be mentioned. Besides Fe_3O_4, iron has two other oxides that are magnetic. Both have the formula Fe_2O_3, but different crystal structures. The more common form is α-Fe_2O_3, which occurs naturally as hematite. It has a trigonal crystal structure, and is weakly magnetic, thought to be a canted antiferromagnet. The other form is γ-Fe_2O_3, called maghemite. This compound has a cubic structure closely related to spinel, with some cation sites vacant to accommodate the lower ratio of iron to oxygen. It is strongly magnetic. Another strongly magnetic compound with important practical applications is CrO_2.

2.4.5 Néel's theory of ferrimagnetism

It has already been mentioned that the properties of ferrimagnetic materials are in many ways similar to those of ferromagnets, but that there are important differences between them. Both classes of materials have a transition temperature below which they exhibit spontaneous magnetization. The spontaneous magnetization varies with temperature, dropping to zero rapidly as the transition temperature is approached from below. At high temperatures, both classes of materials have small positive susceptibilities that decrease with increasing temperature according to eq. (2.2). But, whereas the spontaneous magnetization of ferromagnetic materials has a fairly simple temperature dependence (Fig. 2.1), ferrimagnetic materials exhibit a variety of temperature dependences, which are generally more complicated. Another important difference is that ferromagnets obey eq. (2.2) with a negative sign for θ fairly well at all temperatures above the transition, while in ferrimagnets, the equation is not obeyed near the transition temperature, and θ does not necessarily have a negative sign.

Néel showed that these properties could be explained by assuming that the magnetic moments in ferrimagnets are divided between two sublattices, as in antiferromagnets, but that the magnetic moments on the two sublattices are different either in number or in magnitude. All the main features of the behaviour of ferrimagnetic materials can be derived by applying the molecular field theory to this model. As in the case of antiferromagnets, we assume that the field acting on magnetic moments on the A and B sublattices is given by $H - wM_B$ and $H - wM_A$ respectively, where H is the applied field, w is a constant and M_A and M_B are the magnetizations of the two sublattices. In the case of antiferromagnets, we proceed from here to obtain eqs (2.24) and (2.25), which are symmetrical in M_A and M_B (i.e. if we interchange M_A and M_B, eq. (2.24) turns into eq. (2.25) and vice versa). The corresponding equations for ferrimagnets do not have this symmetry, and this is the reason why ferrimagnets can have a spontaneous magnetization. We will use the simple model in which we take the magnetic moments of all ions to be m, but we assume the number of ions on the A and B sublattices to be in the ratio $\xi{:}\eta$, where $\xi + \eta = 1$. (Note that $\xi = \eta = \frac{1}{2}$ would correspond to an antiferromagnet.) If we let the total number of 'formula units' per unit volume be N, then the same method we have already used to derive eq. (2.7) for paramagnets, eq. (2.11) for ferromagnets and eqs (2.24) and (2.25) for antiferromagnets, will give the following results for ferrimagnets:

$$M_A = 2\xi N m f_c[\mu_0 m(H - wM_B)/kT], \tag{2.38}$$

and

$$M_B = 2\eta N m f_c[\mu_0 m(H - wM_A)/kT]. \tag{2.39}$$

44

As in the case of ferromagnets and antiferromagnets, we cannot solve eqs (2.38) and (2.39) explicitly to give M_A and M_B. But for high temperatures, the function f_c can again be approximated by c times its argument, giving

$$M_A \approx \frac{2c\xi\mu_0 N m^2 (H - wM_B)}{kT}, \qquad (2.40)$$

and

$$M_B \approx \frac{2c\eta\mu_0 N m^2 (H - wM_A)}{kT}. \qquad (2.41)$$

We can solve these equations for M_A and M_B, and we can then obtain an expression for the susceptibility, χ, which is given by $(M_A + M_B)/H$:

$$\chi = \frac{C(T - 2\xi\eta wC)}{T^2 - \xi\eta w^2 C^2}, \qquad (2.42)$$

where

$$C = \frac{2c\mu_0 N m^2}{k}. \qquad (2.43)$$

We can compare eq. (2.42) with equations for χ for other classes of materials (eqs (2.5), (2.8) and (2.31)) more easily if we take the reciprocal:

$$\begin{aligned}
\chi^{-1} &= \frac{T^2 - \xi\eta w^2 C^2}{C(T - 2\xi\eta wC)} \\
&= \frac{T}{C} + 2\xi\eta w - \frac{\xi\eta w^2 C(1 - 4\xi\eta)}{T - 2\xi\eta wC} \\
&= \frac{T}{C} + 2\xi\eta w - \frac{\sigma}{T - \theta}, \qquad (2.44)
\end{aligned}$$

where

$$\sigma = \xi\eta w^2 C(1 - 4\xi\eta) \qquad (2.45)$$

and

$$\theta = 2\xi\eta wC \qquad (2.46)$$

are constants with the dimensions of temperature.

At high temperatures ($T \gg \theta$), the last term in eq. (2.44) can be neglected, and χ^{-1} varies linearly with T (curve 4 in Fig. 2.8). If this linear behaviour were followed at all temperatures, χ would always remain finite, since it would only reach infinity at a temperature θ_p (called the *paramagnetic Curie temperature*) given by

$$\theta_p = -2\xi\eta wC, \qquad (2.47)$$

which is always negative. However, eqs (2.42) and (2.44) show that χ^{-1} departs from linear variation with T as T decreases, and χ becomes infinite at a temperature θ_f (called the *ferrimagnetic Curie temperature*) given by

$$\theta_f = wC(\xi\eta)^{1/2}. \qquad (2.48)$$

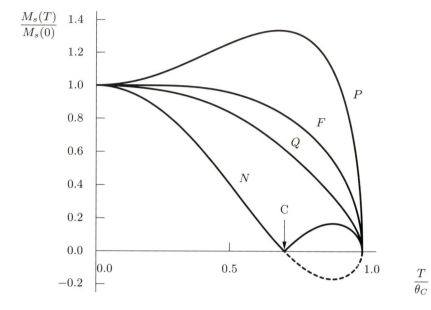

FIG. 2.11   Variation of saturation magnetization with temperature, for P-type, Q-type and N-type ferrimagnets and for ferromagnets (F). C is the compensation point.

Thus, in ferrimagnetic materials, θ_f and θ_p are very different, unlike in ferromagnetic materials, where they are approximately equal. But, as in ferromagnetic materials, θ_f is the critical temperature below which spontaneous magnetization exists. We have seen that in ferromagnetic materials, the spontaneous magnetization obeys a universal temperature variation if plotted in a dimensionless form $(M_s(T)/M_s(0)$ against T/θ_C, Fig. 2.1). Most ferrimagnetic materials depart from the universal curve, although in all cases, $M_s(T)$ becomes approximately constant near $T = 0\,\mathrm{K}$ and drops sharply to zero near $T = \theta_f$. There are three main types of behaviour, which are shown in Fig. 2.11 together with the curve for ferromagnets which is included for comparison. The three types of behaviour can be explained qualitatively as follows. In all ferrimagnets, the magnetic moments on both the A and the B sublattices are perfectly aligned at $T = 0\,\mathrm{K}$, the two sublattices being magnetized in opposite directions. We assume that

$$M_B(0) > M_A(0), \tag{2.49}$$

so that

$$M_s(0) = M_B(0) - M_A(0). \tag{2.50}$$

As T increases, both M_A and M_B decrease. However, in general, M_A and M_B do not decrease with temperature in the same way. In some materials,

M_A decreases faster than M_B. It is easy to see that M_s will decrease relatively slowly at first, and may even reach values greater than $M_s(0)$. These materials are called *P*-type. In other cases, M_B decreases faster than M_A, so that M_s decreases relatively fast, giving the *Q*-type behaviour shown in Fig. 2.11. In more extreme cases (*N*-type materials), where M_B decreases fast relative to M_A, we may reach a temperature at which M_B is just equal in magnitude to M_A. At this temperature, $M_s = 0$, but this is not the Curie temperature, because M_B and M_A are not themselves zero. The temperature at which this happens is called the *compensation point*. As T increases beyond this point, M_A becomes greater than M_B, so that M_s is now equal to $M_A - M_B$, and is directed parallel to M_A. This is indicated by the dashed line in Fig. 2.11. In practice, M_s can only be measured if an external field is applied, to drive the domains (see section 3.2) out of the material. M_s is always parallel to the applied field, so that the observed behaviour follows the solid line rather than the dashed line. (Note that below the compensation point, M_B is parallel and M_A is antiparallel to the applied field, but above the compensation point, M_A and M_B are both reversed). A number of ferrimagnets exhibit compensation points, most notably a number of garnets (but not yttrium iron garnet).

3

Bulk magnetic properties and their measurement

3.1 Magnetization curves

In sections 2.3 and 2.4, we outlined the basic features of ferromagnetic and ferrimagnetic behaviour: the temperature dependence of susceptibility above the transition temperature, and of the spontaneous magnetization below it. We have also mentioned that the magnetization of a ferromagnetic or ferrimagnetic specimen is in general less than the spontaneous value. In order to reach the spontaneous value, it is necessary to apply a magnetic field, though the magnitude of the field needed is always much smaller than that needed to saturate a paramagnet (or a ferromagnet or ferrimagnet above its Curie temperature). In this section, we give a brief, qualitative description of the bulk magnetic properties of ferromagnetic and ferrimagnetic materials. A more detailed explanation of these properties will be given in section 3.2. In these sections, we will be dealing with properties that are common to ferromagnetic and ferrimagnetic materials, and we will therefore refer to them simply as *magnetic materials.*

Although the magnetization of a magnetic specimen in zero field is usually much less than the spontaneous magnetization, it is usually not zero. Most of us have, for example, witnessed the small attraction that can be exerted by a penknife on small iron objects. People who have to handle very small ferromagnetic specimens with tweezers, find it much easier to use tweezers made of a non-magnetic steel, because ordinary steel tweezers sometimes do not let go of the specimens. Mechanical watches and clocks can become inaccurate if they have been near magnets. However, it is possible to demagnetize these objects. This is most easily done by applying an alternating field of amplitude gradually decreasing to zero. A simple demagnetizing device is a coil connected to the a.c. mains. Small objects can be demagnetized by pulling them through the coil. An even better way to demagnetize a specimen is to apply a rotating field of gradually decreasing magnitude. The surest method of all is to heat the specimen above its Curie temperature and cool it down in zero applied field (not recommended for watches and clocks though, nor for any other objects that can be damaged by heating!).

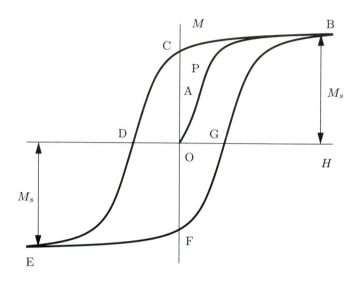

FIG. 3.1 Variation of the magnetization of a typical magnetic material with applied field.

If we apply a magnetic field to a previously demagnetized specimen, its magnetization will gradually increase. The variation of the magnetization, M, in a typical case is illustrated schematically in Fig. 3.1. Starting from point O, the magnetization increases relatively slowly at first, but the increase becomes faster until a maximum rate is reached at A. Beyond A, the magnetization increases more slowly again, until, at B, it reaches a constant value. At this point, the specimen is saturated. The saturation magnetization, M_s, is equal to the spontaneous magnetization discussed in sections 2.3 and 2.4.

If the field, H, is now decreased, M does not retrace the curve B–A–O, but decreases more slowly and when H reaches zero, M still has a non-zero value (point C). In order to decrease M further, we must apply a field in the opposite direction. When this reverse field is sufficiently large, M passes through zero (point D), and then begins to increase in the direction in which H is now applied. Eventually, M reaches the value M_s again (point E). If H is decreased again, M follows the curve from E through F and G back to B again. If we continue to change H between large values in opposite directions, M will vary repeatedly along the closed loop B → C → D → E → F → G → B. This loop is called the *hysteresis loop*, whereas the curve O → A → B is the *initial magnetization curve*. The shapes of these curves can be very different in different materials. For different practical applications, we need materials with magnetization curves of different shapes. In this section, we outline the main features of the curves that can be used to characterize them. In the following section, we will discuss the reasons for hysteresis, and ways to control the shapes of magnetization curves to produce useful materials.

In diamagnetic, paramagnetic and antiferromagnetic materials, M is proportional to H and therefore the susceptibility, χ, is independent of H. Therefore, χ has an unambiguous meaning in these materials. In ferromagnetic and ferrimagnetic materials, the relationship between M and H is very complex, and we must therefore be very careful in defining what we mean by susceptibility. In fact, the susceptibility has a clearly defined meaning above the Curie temperature, as shown in eqs (2.19) and (2.42). But as both these equations show, χ becomes infinite at the Curie temperature, θ_C. If we want to extend the concept of susceptibility to below θ_C, it will have to be given a different meaning. Above θ_C, χ is the ratio of the *local* magnetization and the applied field. Below θ_C, the *local* magnetization is equal to M_s even in zero applied field. However, the local magnetization can vary in direction from one part of the specimen to another, and therefore the *average* magnetization is usually less than the local magnetization. What is plotted in Fig. 3.1 is the average, not the local magnetization. (The two become equal at points B and E.) We can therefore define a susceptibility as being some relationship between the average magnetization and H. But even here we must be careful. At points C and F, the ratio M/H is infinite, and along the segments CD and FG, M/H is negative. The susceptibility is evidently not a very useful concept along the hysteresis loop. Along the initial magnetization curve, M/H is always finite and positive, even though it varies in magnitude with H (or M). The ratio M/H is defined as the *total susceptibility*, χ_t. Alternatively, we can use the rate of change of M with H, i.e. dM/dH as the susceptibility. This is called the *differential susceptibility*, χ_d. Both susceptibilities tend to 0 in high fields, although, as can be seen from Fig. 3.1, χ_t decreases more slowly than χ_d. It is instructive to plot these susceptibilities as a function of M rather than H. These plots are shown schematically in Fig. 3.2. Both χ_d and χ_t pass through maxima. The maximum of χ_d corresponds to the point of inflection, A, in Fig. 3.1, whereas the maximum of χ_t corresponds to the 'knee' of the curve, P. We see that the maximum of χ_d is greater, and occurs at a lower value of M, than that of χ_t. The two susceptibilities are equal near $M = 0$. The value at this point is called the *initial susceptibility*, χ_a. The value of χ_a, as well as the maxima of χ_d and χ_t, are often used to characterise magnetic materials, particularly those intended for use as transformer cores.

We have already seen that if the magnetic field varies between large values in opposite directions, the magnetization varies in a complicated, irreversible way. However, irreversible effects occur even for small changes of field. Consider what happens if we vary the magnetic field, starting from the point O in Fig. 3.1. If we apply a very small field and then remove it, the magnetization will return to zero. If we increase the field to a larger value,

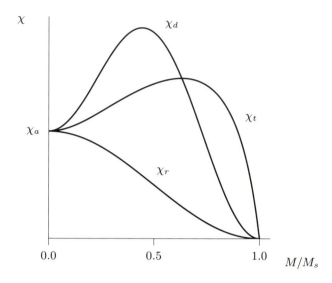

FIG. 3.2   Variation of various susceptibilities with M.

and then decrease it slightly, the magnetization will not retrace the original curve towards O. The behaviour is shown schematically in Fig. 3.3. If we increase the field successively to points Q, R, S and T, and at each of these points, we stop and take the field through a small cycle, the magnetization describes a small loop in each case. If we gradually decrease the amount by which we change the field, the loops become smaller and narrower, and tend eventually to be just small straight lines. The gradients of these lines are also equivalent to a type of susceptibility, called the *reversible susceptibility*, χ_r, plotted in Fig. 3.2. We see that χ_r is also equal to χ_a at $M = 0$, but as M increases, χ_r decreases steadily to zero.

A few important parameters are used to describe the hysteresis loop. The largest magnetization we can get in zero field, at points C and F, is called the *remanent magnetization* or *remanence*, M_r. The reverse field needed to bring the magnetization to zero from remanence, points D and G, is called the *coercive field* or *coercivity*, H_c. The area of the loop is also an important parameter. Each time the complete loop is traversed, an amount of energy equal to μ_0 times the area is dissipated.

Life would be simple (but not so interesting) if we could now make a list of properties that would make a material 'the best magnetic material of all'. But a set of properties that might make a magnetic material very good for a particular application, would render it completely unsuitable for another. Just about the only rule we can make is that M_s should be as large as possible—and even that rule is not always right (see section 4.2.4.5). As for the other parameters, such as μ, M_r, H_c, etc., large values are useful

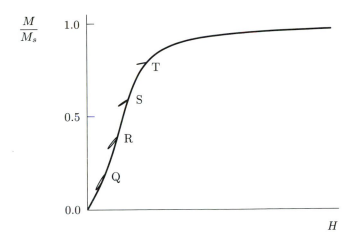

FIG. 3.3   The initial magnetization curve, showing minor loops.

sometimes and small values at other times. These points will be discussed later. Our immediate task now is to gain some understanding of why the magnetization of magnetic materials behaves as it does. This is the subject of the next section.

3.2   Magnetic domains

It has already been mentioned that magnetic materials were generally (except in reasonably large applied magnetic fields) subdivided into domains. Within each domain, the magnetization is uniform and equal to M_s, but different domains are magnetized in different directions. The average magnetization of a specimen could therefore be much less than M_s, and could even be zero. In this section, we will explain, at least qualitatively, why a magnetic material prefers to be subdivided into domains rather than be uniformly magnetized. The reason for this can be stated very simply: a material will always be in a state in which its energy is a minimum. What we have to explain therefore is why a subdivided state has a smaller energy than a uniformly magnetized state. After all, the reason why the material is magnetic in the first place is that there are strong forces tending to align all the magnetic moments in the same direction. Without these forces, the material would be paramagnetic, not ferromagnetic.

3.2.1   Exchange energy

In section 2.3.2 we introduced the idea of an internal field, which is responsible for producing the aligning forces. But we also saw that the internal field concept is not entirely satisfactory: it predicts that the Curie temperature, θ_C, should be equal to the constant, θ, in eq. (2.8), whereas these two temperatures are not quite equal. We explained this by saying that the internal field is not really proportional to the average magnetization, but to the local magnetization. In fact, the internal field is produced by interactions between magnetic moments that are near neighbours—in many cases, only the interactions between *nearest* neighbours are significant.

Consider two neighbouring atoms whose magnetic moments can be represented by the vectors $\mathbf{m}_1$ and $\mathbf{m}_2$. (The magnetic moments have a direction as well as magnitude, so they must be represented as vectors. Vector notation is explained in appendix C.) The two magnetic moments have a potential energy, W_e, which depends on their relative orientation:

$$W_e = -2\beta|\mathbf{m}_1||\mathbf{m}_2|\cos\theta, \tag{3.1}$$

where β is a constant and θ is the angle between the vectors $\mathbf{m}_1$ and $\mathbf{m}_2$. The factor 2 is included here in order to simplify some of the equations that will be derived later. The reason for the subscript e is that it stands for *exchange* which is a term originating from quantum mechanics. To calculate the magnitude of β from first principles we would need to use quantum mechanics, but, as many readers will no doubt be thankful to learn, we shall be satisfied with saying that β is just a constant. In fact, it is obvious from eq. (3.1) that the condition for ferromagnetism is that β should be positive, because if that is so, then W_e is a constant and is a minimum value when $\theta = 0$, i.e. when $\mathbf{m}_1$ and $\mathbf{m}_2$ point in the same direction. Eq. (3.1) can be written more conveniently, using the notation for scalar multiplication of vectors, as

$$W_e = -2\beta\mathbf{m}_1.\,\mathbf{m}_2, \tag{3.2}$$

but, as we can take $|\mathbf{m}_1| = |\mathbf{m}_2|$, eqs (3.1) or (3.2) can be written

$$W_e = -2\beta m^2\cos\theta, \tag{3.3}$$

where we have written m for the magnitudes of the vectors $\mathbf{m}_1$ and $\mathbf{m}_2$. We can therefore express the exchange energy per unit volume of a material as

$$E_e = -2\beta m^2 \sum_i \sum_j \cos\theta_{ij}, \tag{3.4}$$

where the summation is over all nearest neighbour pairs, i and j, in a unit volume. (Note that in this section, the symbol W is used to denote energy, and E to denote energy per unit volume.)

54

Consider now what happens if the magnetic moments in the material are not parallel to each other. There will then be an increase in E_e, but this increase will be small provided that all θ_{ij}-s are small. If they are small, then we can use the approximation

$$\cos\theta_{ij} \approx 1 - \tfrac{1}{2}\theta_{ij}^2, \tag{3.5}$$

so that eq. (3.4) becomes

$$E_e = C + \beta m^2 \sum_i \sum_j \theta_{ij}^2, \tag{3.6}$$

where C is a constant, which is of no interest, because it is unaffected by the direction of the magnetic moments, and therefore it contributes nothing to any *change* in E_e. We will therefore officially declare C to be zero from now on. E_e is obviously smallest if all θ_{ij}-s are zero, i.e. if all magnetic moments are parallel to each other. Now, if there were no other type of energy to consider, then obviously all magnetic materials would be magnetized to saturation all the time. We know that this is not so, and therefore it would be useful at this stage to calculate E_e for a simple case of non-uniform magnetization. We may not realise yet why the magnetization should be non-uniform, but the result will be useful later.

Let us consider a material with a simple cubic crystal structure with a nearest-neighbour interatomic distance a. Let the x, y and z directions be parallel to the cube edges. We assume that the magnetic moments are everywhere parallel to the yz plane. The orientation of the magnetic moments varies with x so that the angle between neighbouring moments is $\delta\theta$ (see Fig. 3.4). In a unit volume of material, the number of pairs of neighbouring magnetic moments at an angle $\delta\theta$ to each other is $1/a^3$. Hence

$$E_e = \frac{\beta m^2}{a^3}(\delta\theta)^2. \tag{3.7}$$

But as $\delta\theta$ is small, we could think of the angle θ (the angle between the magnetic moments and the y axis) as varying continuously with x, rather than changing by a finite amount $\delta\theta$ after each step of length a. We can therefore write

$$\delta\theta \approx a\frac{d\theta}{dx}, \tag{3.8}$$

so that

$$E_e = \frac{\beta m^2}{a}\left(\frac{d\theta}{dx}\right)^2, \tag{3.9}$$

or

$$E_e = A\left(\frac{d\theta}{dx}\right)^2, \tag{3.10}$$

where A is called the *exchange constant*. Eq. (3.10) applies to materials with any crystal structure, but A is equal to $\beta m^2/a$ only for simple cubic

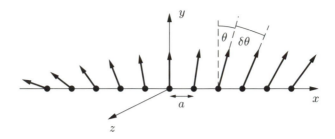

Fɪɢ. 3.4 Illustrating the rotation of the magnetic moments of atoms along the
x axis.

crystals. In other crystal structures, A is some constant times $\beta m^2/a$. Note
that A has dimensions of energy/length, and its unit in the SI system is
$\mathrm{J\,m^{-1}}$.

3.2.2 Demagnetizing energy

We now come to the question of why a magnetic material should not always
be uniformly magnetized. Clearly, the exchange energy is not the only type
of energy that a magnetic material has. Fig. 3.5 illustrates how a second
type of energy arises. A uniformly magnetized specimen (Fig. 3.5(a)) generates a large amount of stray magnetic field, H. This field has an energy
that can be expressed as

$$W_d = \tfrac{1}{2}\int H^2 dV, \tag{3.11}$$

where the integral is to be evaluated over all space. Now, calculating H
and then working out the integral may not be a simple matter even for a
simple magnetization configuration such as that of Fig. 3.5(a). But we can
easily guess that W_d will be much smaller for a specimen subdivided into
two domains (Fig. 3.5(b)), and will get gradually less as the specimen is
subdivided into more and more domains, simply because H is confined to
a smaller and smaller region near the specimen surface. There is another
way to understand why the uniformly magnetized state is unstable. In
Fig. 3.5(a), the specimen carries magnetic north poles on its top surface,
and south poles on its bottom surface. The free poles are formed on any
surface along which there is a discontinuous change in the component of
magnetization normal to the surface. These poles can be thought of as
the sources of H, which always points towards south poles and away from
north poles. A field is therefore also present *inside* the specimen, where
it points in the opposite direction to the magnetization. (This field is not
shown in the figure to avoid confusion.) As the magnetization is in the
opposite direction to the field, the situation is obviously unstable. It is

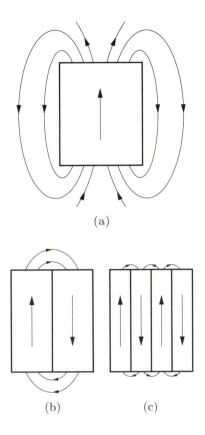

(a)

(b)   (c)

FIG. 3.5   The stray field of (a) a uniformly magnetized specimen, and of a specimen divided into (b) two, and (c) four domains.

no good merely to reverse the magnetization, because then H is reversed
as well, and the configuration is still unstable. For this reason, the field
inside the specimen is called the *demagnetizing* field, and this is why the
subscript d is used on the left-hand side of eq. (3.11). In Fig. 3.5(b), the top
surface of the specimen carries north poles on the left and south poles on
the right, and the bottom surface carries south poles on the left and north
poles on the right. The demagnetizing field does not extend from the top
surface to the bottom, but it is confined to the region near the two ends
of the specimen. Again, the field inside the specimen is not shown, but it
can be seen that its main effect is to try to turn the magnetization to be
parallel to the end surfaces, but it only has this effect near the ends—over
most of the volume of the specimen, the magnetization is not affected by
the demagnetizing field. As the specimen is further subdivided into smaller
domains, the effect of the demagnetizing field becomes even smaller.

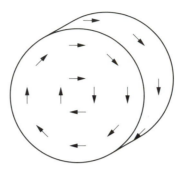

FIG. 3.6   A magnetization configuration that has very little exchange and demagnetizing energy.

3.2.3   Anisotropy energy

We can now understand, at least qualitatively, why magnetic materials are
not usually uniformly magnetized to saturation. Although the exchange
energy would be a minimum for the saturated state, the sum of the exchange plus demagnetizing energies would not be. If that is so, then what
is wrong with a configuration such as that shown in Fig. 3.6, in which, for
simplicity, we have represented the specimen as cylindrical? It has no free
magnetic poles, therefore no demagnetizing energy, and as the magnetization rotates slowly everywhere, it has very little exchange energy. (The
observant reader may be wondering what is happening along the axis of the
cylinder. To avoid the inevitable complications, a hollow cylinder could be
considered.) Yet this kind of configuration does not normally exist—we are
much more likely to find something like Fig. 3.5(c), or a more complicated
version of it. In other words, there is usually a structure consisting of uniformly magnetized domains, separated by narrow boundaries. The reason
(it might not be difficult to guess by now) is that there is another kind
of energy we have not considered yet. It is called *anisotropy*, and it arises
from the crystalline nature of most magnetic materials. The anisotropy energy has been demonstrated by measuring magnetization curves in single
crystal specimens. Fig. 3.7 shows magnetization curves for iron (which has
a body-centred cubic crystal structure) with the field applied parallel to
[100], [110] and [111] directions respectively. Evidently, it is much easier to
magnetize iron in the [100] direction than in the [110] or [111] directions.
Similar measurements for nickel (face-centred cubic) show that the easiest
magnetization direction is [111], whereas in cobalt (close-packed hexagonal) it is [0001]. These results suggest that the anisotropy energy depends
on the direction of magnetization relative to the crystal lattice.

   In a cubic crystal, we usually express the direction of magnetization in

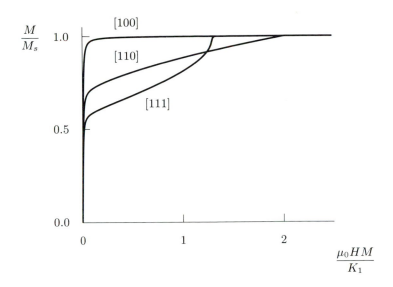

FIG. 3.7   Magnetization curves for a single crystal of iron with the field applied parallel to various crystallographic directions.

terms of the cosines of the angles it makes with each of the cube edge directions. These *direction cosines* are denoted by α_1, α_2 and α_3, which are always related by

$$\alpha_1^2 + \alpha_2^2 + \alpha_3^2 = 1. \tag{3.12}$$

The anisotropy energy per unit volume, E_a, is a function of α_1, α_2 and α_3. It is usual to express E_a as a series in increasing powers of the α-s. We can then usually neglect all but the lowest one or two powers. For a cubic crystal, the expression for E_a is simplified because of the high degree of symmetry of the crystal. As the energy must be unchanged if the magnetization direction is reversed, E_a can only depend on even powers of the α-s. The symmetry of the crystal also requires that E_a should be unchanged if the α-s are interchanged with each other. For example, E_a in a crystal magnetized in the [100] direction must be the same as that in a crystal in the [010] or [001] directions. The lowest power term that satisfies these requirements is the term proportional to $\alpha_1^2 + \alpha_2^2 + \alpha_3^2$, but, from eq. (3.12), this term is constant. The lowest power we need to consider therefore is the fourth, and the next is the sixth. We can therefore write

$$E_a = K_1(\alpha_1^2\alpha_2^2 + \alpha_2^2\alpha_3^2 + \alpha_3^2\alpha_1^2) + K_2(\alpha_1^2\alpha_2^2\alpha_3^2) + \cdots, \tag{3.13}$$

where K_1 and K_2 are constants. They have different values for different materials, and may even vary with temperature in the same material. In the first approximation, let us neglect the K_2 term in eq. (3.13). The two extreme values of E_a will then be 0 when the magnetization is parallel to the [100] direction, say ($\alpha_1 = 1$, $\alpha_2 = \alpha_3 = 0$), and $\frac{1}{3}K_1$ when the magnetization is parallel to [111] say ($\alpha_1 = \alpha_2 = \alpha_3 = 1/\sqrt{3}$). Therefore, if K_1 is positive, then E_a is a minimum when the magnetization is parallel to [100] (as in iron), but if K_1 is negative, then E_a is a minimum when the magnetization is parallel to [111] (as in nickel).

In cobalt, as well as in other materials with hexagonal or tetragonal crystal structures (these crystals are known as *uniaxial*), E_a is determined mainly by the angle ϕ between the magnetization and the main symmetry axis. We can express E_a as a series of powers of $\sin\phi$, and again, we need only consider even powers. Therefore

$$E_a = K_1 \sin^2\phi + K_2 \sin^4\phi + \cdots, \tag{3.14}$$

where K_1 and K_2 are again constants. Neglecting K_2 again, E_a has the extreme values 0 when $\phi = 0$ and K_1 when $\phi = \frac{1}{2}\pi$. Again, it depends on the sign of K_1 which of these values is the minimum. In cobalt, at room temperature, K_1 is positive and therefore E_a is a minimum when $\phi = 0$, i.e. when the magnetization is parallel to [0001]. But if K_1 is negative, then the magnetization will prefer to be at right angles to [0001], i.e. it will lie in one of the directions parallel to the basal plane. This happens in cobalt above about 340°C.

3.2.4 Energy and width of domain walls

We can now understand the formation of domains qualitatively. A configuration such as that shown in Fig. 3.6 would have too much anisotropy energy, because in a large part of the specimen, the magnetization is not parallel to an 'easy' direction, whereas configurations such as those shown in Fig. 3.5 would have very small anisotropy energies if the vertical direction happened to be an easy direction. The domain structure can be even better understood if we consider the boundary between two domains. The energy of this domain wall can be estimated from the model used to derive eq. (3.7) (see Fig. 3.4). We assume that the magnetization rotates by an angle $\delta\theta$ in a distance a, until θ reaches values of $\pm\frac{1}{2}\pi$ on either side of the plane $x = 0$, and beyond that, it remains constant. The region in which θ varies is the domain wall. As in this case the wall separates two domains magnetized in opposite directions, we describe it as a 180° wall. We denote the width of the wall by w_{wall}. If the wall extends over N interatomic distances, then

$$w_{\text{wall}} = Na, \tag{3.15}$$

and

$$\delta\theta = \frac{\pi}{N}. \tag{3.16}$$

Using eqs (3.7) and (3.16) and writing $A = \beta m^2/a$, the exchange energy per unit volume inside the domain wall is

$$E_e = \frac{A\pi^2}{N^2 a^2}. \tag{3.17}$$

For a unit area of domain wall, the exchange energy is

$$\gamma_e = \frac{A\pi^2}{Na}. \tag{3.18}$$

We assume that the two domains are magnetized in easy directions, so that there is no anisotropy energy associated with the domains. There is however some anisotropy energy associated with the wall, since the magnetization in the wall is in general not parallel to an easy direction. As a rough approximation, we may assume that the anisotropy energy per unit area of wall is

$$\gamma_a = KNa, \tag{3.19}$$

where K is the anisotropy constant. The width of the wall, i.e. the value of N, will adjust itself to make the total energy per unit area,

$$\gamma = \gamma_e + \gamma_a = \frac{A\pi^2}{Na} + KNa, \tag{3.20}$$

a minimum. Hence we must have $d\gamma/dN = 0$, which gives

$$-\frac{A\pi^2}{N^2 a} + Ka = 0, \tag{3.21}$$

or

$$N = \frac{\pi}{a}\left(\frac{A}{K}\right)^{1/2}. \tag{3.22}$$

The wall width is therefore

$$w_{\text{wall}} = Na = \pi\left(\frac{A}{K}\right)^{1/2}. \tag{3.23}$$

Substituting eq. (3.22) into eq. (3.20), we can calculate the wall energy per unit area:

$$\gamma = 2\pi(AK)^{1/2}. \tag{3.24}$$

In a typical magnetic material, we may find that $A \approx 10^{-11}\,\text{J}\,\text{m}^{-1}$ and K is of the order of 10^3 to $10^5\,\text{J}\,\text{m}^{-3}$. This shows that w_{wall} is of the order of 100 nm, which is a few hundred interatomic distances. Now, the observed size of domains is usually much larger than this value, which explains why magnetic materials are subdivided into domains within which the magnetization is uniform, rather than the magnetization rotating slowly throughout the specimen as in Fig. 3.6.

With the above values for A and K, we can estimate γ to be of the order of $10^{-3}\,\text{J}\,\text{m}^{-2}$. It must be emphasised that this domain wall energy is not a separate type of energy. All that we have shown is that the exchange and anisotropy energies are normally concentrated in the domain walls.

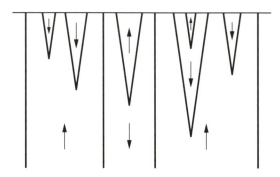

FIG. 3.8   Reverse spike domains near the surface of a uniaxial crystal.

3.2.5   Magnetostriction

We have now considered all the energy terms needed to explain the domain structure of uniaxial materials such as cobalt. The structure will consist of long, narrow domains magnetized in opposite directions, rather like Fig. 3.5(c). Near surfaces or grain boundaries, 'reverse spike' domains often form (Fig. 3.8). Their role is to decrease the demagnetizing energy by introducing small areas of poles of opposite sign to those on the ends of the 'main' domains, without adding too much extra domain wall energy. There is however one more type of energy needed to explain the domain structure of cubic materials. Consider a cubic single crystal in the shape of a cube whose surfaces are parallel to easy directions. Fig. 3.9 shows two possible domain structures for such a crystal. Neither structure contains any free poles, and therefore neither has any demagnetizing energy. As the structure in Fig. 3.9(a) contains fewer domain walls, it has less domain wall energy, and therefore it should be more likely to occur than the more complicated structure of Fig. 3.9(b). Yet structures as simple as that of Fig. 3.9(a) do not usually occur. To explain the domain structure of cubic materials, we have to consider the interaction between the magnetization and elastic distortions.

When the magnetization of a specimen is changed, there is a slight change in its shape, generally of the order of 1 part in 10^5 or less. Some materials expand in the direction of the magnetization, others (notably nickel) contract. This effect is called *magnetostriction*; the former type of materials are said to have positive magnetostriction, the latter type negative. There is also an inverse effect: if a tensile stress is applied to a material with positive magnetostriction, it will be easier to magnetize it parallel and harder to magnetize it perpendicular to the axis of the tension, whereas if a compressive stress is applied, then it will be harder to magne-

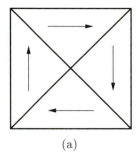
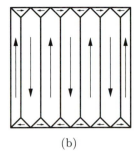

(a)   (b)

FIG. 3.9   Two possible domain structures in a cubic single crystal.

tize it parallel to the axis of the compression. For a material with negative magnetostriction, tension and compression have the opposite effect. The material effectively acquires an extra contribution to its anisotropy, induced by the stress.

The distortion associated with the magnetization is always present in the specimen. The domains are magnetized to saturation, and therefore each domain is distorted. A demagnetized specimen contains domains magnetized in various directions, and the different domains are distorted in different directions. When the specimen is magnetized, the domain magnetizations are aligned in the direction of the field, and the axes of the distortions are also aligned. This is what causes the specimen to change its shape. Consider now the elastic distortions associated with the domain structure of Fig. 3.9(a). Fig. 3.10 shows the way in which the material tries to be deformed, depending on whether the magnetostriction is (a) positive or (b) negative. But magnetic materials are not usually observed to fall apart spontaneously. In order to prevent the specimen from splitting along the domain walls, the distortion of the domains must be counteracted. This requires extra elastic energy. The specimen of Fig. 3.9(b) can however be freely deformed over most of its length, since the distortion produced by magnetization in opposite directions is the same. Some elastic energy is needed to hold the small triangular domains (called *closure domains*) to the rest of the specimen, but this energy is much less than that required for the large domains of Fig. 3.9(a). The size of the domains in cubic materials is therefore generally limited by the magnetoelastic energy.

By considering the four types of energy discussed above, we can explain at least qualitatively the subdivision of magnetic materials into domains, and we can even make a reasonable guess about the arrangement of domains in a particular specimen. We cannot usually go further than that—the exact structure is not only very difficult, if not impossible to calculate, but there can be many different domain configurations that can be stable in the same specimen. As can be seen from Fig. 3.1, the magnetization

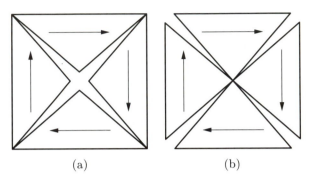

FIG. 3.10   Possible elastic distortions in materials with (a) positive and (b) negative magnetostriction.

in zero field can have any value lying on the M-axis between the points C and F. These different values of M obviously correspond to different domain structures, but even one particular value of M can be achieved with different domain structures.

It is interesting to note that when Weiss postulated the existence of magnetic domains (see section 2.3.2), he was merely aiming to explain the apparent contradiction between the small magnetization observed in most ferromagnets and the large spontaneous value predicted by the theory. The original Weiss theory did not include any explanation of *why* magnetic domains existed, it merely suggested that *if* they existed, they would account for the observed magnetic properties of ferromagnetic materials. (Similarly, the Weiss hypothesis does not extend to predicting the size and shape of domains.) The reason for the existence of domains was understood only in the 1930s when the importance of the demagnetizing energy was recognised. However, at about the same time, magnetic domains were also observed experimentally, so that we now have not only a theoretical justification for their existence, but also experimental proof. Methods of observing magnetic domains will be described in section 3.3.5.

3.2.6   The effect of applied fields

When a magnetic field is applied, the magnetization changes, and so, obviously, the domain structure changes too. The reason for the change can be stated in terms of the energy due to the applied field. This energy, per unit volume is

$$E_H = -\mathbf{H}.\mathbf{M}. \tag{3.25}$$

For a given H, this is a minimum when H and M are parallel, and a maximum when they are antiparallel. Fig. 3.11, which shows a small region somewhere inside a specimen, illustrates the effect of an applied field. We

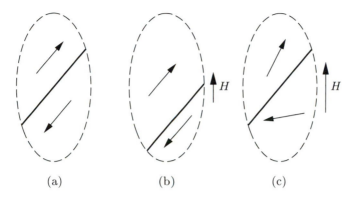

FIG. 3.11   Possible effects of an applied field on a region inside a specimen, (a) no field applied, (b) domain wall motion in a small field, and (c) magnetization rotation in a larger field.

have chosen a region containing parts of two domains separated by a wall. When H is applied, E_H is greater in the domain on the right than in the domain on the left. As H increases, this energy difference increases. The energy can be reduced in two ways. Firstly, the domain wall can move, as in Fig. 3.11(b), thereby increasing the volume of the domain whose energy is lower and decreasing the volume of the other. Secondly, the magnetization direction of the two domains can change, as in Fig. 3.11(c). Both these processes can occur in practice.

The position of the domain walls depends on the demagnetizing energy. When an external field is applied, the field inside the specimen is the sum of the demagnetizing field and the applied field. The domain walls will therefore generally move to new positions. On the other hand, the magnetization directions in the domain are determined mainly by the anisotropy, which resists the rotation of magnetization. In general therefore, domain wall motion tends to occur in small fields, and magnetization rotation only begins as the field reaches larger values—in some cases, rotation only starts when the field is large enough to have driven all domain walls out of the specimen. But there are exceptional cases. Consider a specimen with the type of domain structure shown in Fig. 3.5(c). If a field is applied parallel to the magnetization of one set of domains, then alternate domain walls move in opposite directions until pairs of them meet and are annihilated. The specimen can therefore reach saturation purely by domain wall motion. However, if the field is applied perpendicular to the magnetization directions, then there is no difference in the energy of the two sets of domains, so the domain walls do not move. The magnetization directions gradually rotate towards the direction of the applied field until the specimen becomes uniformly magnetized. Thus, saturation is reached entirely by magnetization rotation.

These two extreme situations do occur in some practical materials. If a wire made from an alloy of 78% Ni, 22% Fe (one of the range of alloys with the trade name *Permalloy*) is heat treated in a magnetic field parallel to the length of the wire, it acquires an easy direction of magnetization parallel to this field. Subsequently, the domain structure will consist of long, narrow domains magnetized in one of the two directions parallel to the wire. If a field is applied along one of these directions, it causes the domain walls to move until the wire is saturated. The opposite situation may be achieved in strips of a 50% Ni, 50% Fe alloy by cold rolling. The easy directions are then in the plane of the strip, but perpendicular to its length. If a field is now applied parallel to the length of the strip, there will be no domain wall motion, but saturation will be reached entirely by the rotation of the magnetization towards the direction of the applied field. During this process, the magnetization changes linearly with the field, so that the permeability remains constant for a fairly wide range of applied fields. For this reason, the material has been given the trade name *Isoperm*. (Permalloy and Isoperm materials will be discussed again in section 4.1.3.2.)

It is more usual however for both domain wall motion and magnetization rotation to occur. In general, both processes may occur reversibly or irreversibly.

3.2.7   Domain wall motion

In many materials, domain walls move reversibly in very small applied fields. In other words, the walls are displaced by a small amount when the field is applied, but if the field is removed, they return to their original positions. In larger fields, the motion becomes irreversible—the walls do not return to their original positions when the field is removed. The main reason is that the energy of the domain walls is not constant, but varies in an irregular manner because of inhomogeneities in the specimen. Inhomogeneities that can cause fluctuations of wall energy are inclusions of a second phase, dislocations, grain boundaries, internal stresses, groups of point defects or impurity atoms, voids, etc. For a domain wall in the yz plane, able to move in the x-direction, the wall energy may vary as shown schematically in Fig. 3.12. Initially, the wall is at A in zero field. In a small applied field, it moves by a small distance, to B, say. If the field is removed, it will return to A. However, if the field is large enough to move the wall to C, where the slope of the $\gamma(x)$ curve is largest (i.e. where there is a point of inflection), the wall can continue to move without any further increase of field. It will come to rest at F, where the slope of the $\gamma(x)$ curve is the same as at C. (Note that this spontaneous jump occurs at C, where the *slope* is a maximum and not at D, where γ is a maximum. This is because the force on the wall is proportional to the slope, $d\gamma/dx$.) If the field is now removed, the wall will not return to A, but to E. Thus, the jump from C

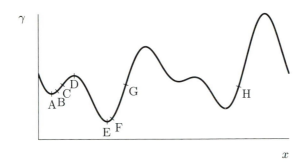

FIG. 3.12   Variation of domain wall energy with position of the wall in the specimen.

to F is irreversible—a field in the opposite direction is needed for the wall
to return to A. If, on the other hand, the field is increased further after the
wall has reached F, it will move towards G, and from there, it will jump to
the next upward slope, at H. In this way, we can illustrate irreversible wall
motion in a simple way. In reality, the situation is more complicated. When
a wall moves, it encounters obstacles distributed randomly throughout the
specimen, and therefore parts of the wall are retarded while other parts
bulge forward. It is therefore not usually possible to represent the wall
energy as a simple function of one variable (x), or even of three variables
(x, y, z). Irreversible wall motion is thus very difficult to treat theoretically.
Nevertheless, it should be appreciated that the way to make it easier for
domain walls to move easily in a material is to eliminate all inhomogeneities
as far as possible.

3.2.8   Magnetization rotation

Magnetization rotation—both reversible and irreversible—is easier to treat
quantitatively. For the case in which the magnetization rotates *coherently*
(i.e. it remains uniform even while it is rotating), the theory is well understood. The simplest case is that of a small particle with a positive uniaxial
anisotropy. The anisotropy energy of such a particle can be written

$$E_a = KV \sin^2 \phi, \tag{3.26}$$

where K is a constant (assumed to be positive), V is the volume of the
particle, and ϕ is the angle between the magnetization and some fixed
direction in the particle. The anisotropy may arise from various causes: it
may be due to a hexagonal or tetragonal crystal structure, or to a non-
spherical shape, or to stresses. These factors determine the value of K, but
the dependence on ϕ will always be according to eq. (3.26) if the anisotropy
is uniaxial. To make the effect easy to illustrate, we consider a particle in

67

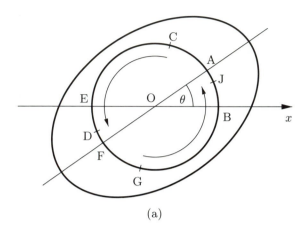

(a)

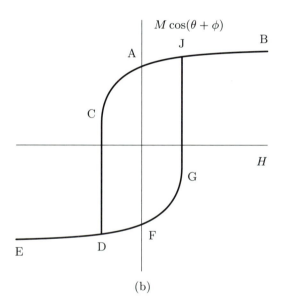

(b)

FIG. 3.13   (a) Rotation of the magnetization of a spheroidal particle in an applied field. (b) The resulting magnetization curve.

the shape of a prolate spheroid, as shown in Fig. 3.13(a). Such a shape leads to a uniaxial anisotropy with easy directions parallel to the major axis. The value of K depends on the ratio of the major to the minor axis of the spheroid.

When there is no magnetic field applied, the magnetization points in the direction OA—one of the easy directions. If a gradually increasing field is applied along Ox, the magnetization will gradually rotate towards OB. In Fig. 3.13(b), the magnetization is plotted as a function of field, or rather, the component of the magnetization along Ox, which is given by $M_s \cos(\theta + \phi)$, where θ is the angle between the easy axis and Ox. (Note that ϕ is measured *clockwise* from the easy direction OA in Fig. 3.13(a)). Corresponding points are denoted by the same letter in Fig. 3.13(a) and (b). The point B corresponds to saturation in the direction of the field. If the field is reduced, the magnetization returns to A. Suppose we now apply a field in the $-Ox$ direction. The magnetization rotates to C, say. At C, the magnetization suddenly jumps to a new direction, D, because at this point, the field has become large enough to pull the magnetization from near one easy direction (A) to near the other (F). If the field is increased further, then eventually saturation is reached, at E. If the field is now reduced, the magnetization rotates from E to D, and reaches F when the field reaches zero. If an increasing field is now applied in the $+Ox$ direction, the magnetization gradually rotates to G, from where it suddenly jumps to J. On further increase of the field, the magnetization rotates to B, and the cycle is then repeated as the field is changed from a large value in the $+Ox$ direction to a large value in the $-Ox$ direction and back again. Thus, this process leads to hysteresis, as shown in Fig. 3.13(b). The shape of the hysteresis curve depends on the angle θ, and curves for various values of θ are shown in Fig. 3.14. For $\theta = 0$, the curve is rectangular, and the coercivity is largest. As θ increases, the coercivity decreases, and for $\theta = 90°$, it is zero. The largest coercivity, at $\theta = 0$, can be shown to be $2K/\mu_0 M_s$. In the case when the anisotropy is due to the crystal structure, this maximum coercivity is referred to as the *anisotropy field*, because it is the field needed to rotate the magnetization from an easy to a hard direction.

The curves of Fig. 3.14 are fairly easy to calculate—although the relevant equations cannot be solved analytically, the numerical computations are simple enough to be done on a microcomputer in a few minutes. The theory of magnetization rotation is therefore relatively simple, compared with that of domain wall motion. It is a pleasant surprise that such a relatively simple theory can predict hysteresis! Unfortunately, things do not work out so simply in practice. First of all, a collection of small particles usually includes many different orientations, so that we have to average over a range of values of θ. Secondly, the coercivity is in most cases found to be smaller than the value predicted by the theory. There are various possible reasons for this. For example, in a real material, the magnetization does not usually rotate coherently, and at non-zero temperatures, thermal agitation can help to reverse the magnetization. Nevertheless, as we shall see later, the theory of magnetization rotation has inspired a search for new useful materials that has had many successes.

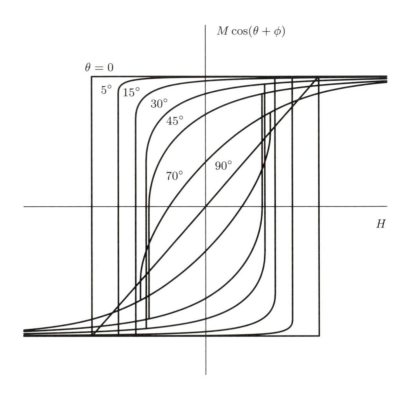

F IG. 3.14   Magnetization curves for various values of the angle θ between the direction of the applied field and the easy axis, for coherent magnetization rotation.

3.3   Measurement of magnetic properties

As we have seen, the most important magnetic parameters that characterise a magnetic material are those related to its hysteresis loop. If we want to determine the usefulness of a material for a particular application, we need to measure its hysteresis loop, i.e. its magnetization as a function of applied field. From these measurements, we can usually deduce values of the saturation magnetization, coercivity, remanence, $U_{\max}$, and permeability or susceptibility. We may also need to know how some of these properties vary with temperature. For example, the Curie temperature can be deduced from measurements of the variation of M_s with temperature. Magnetic measurements are of interest not only in cases where the magnetic properties themselves are to be exploited, but they often have a more fundamental importance. The susceptibility of paramagnetic materials can also give important information about the structure of materials such as multiphase alloys or about defects in materials.

In addition to magnetization and susceptibility, other fundamental properties such as anisotropy and magnetostriction are also of interest in many cases.

In this section, we briefly discuss methods for measuring magnetization, magnetic susceptibility, the Curie temperature, anisotropy and magnetostriction. We then describe methods of imaging magnetic domains. Finally, we mention some examples of using magnetic measurements in metallurgical studies.

3.3.1   Measurement of magnetization and susceptibility

The methods of measuring magnetization fall into two categories. In the first category are methods that detect a current induced in a circuit placed close to a magnetic material. In the second category are measurements of the force on a magnetic specimen in an inhomogeneous field. Usually, it is required to measure the magnetization as a function of an applied field, and this can cause problems because of the presence of demagnetizing fields—the field acting on the specimen is the sum of the applied field and the demagnetizing field, and as these two fields are usually in opposite directions, the effective field can be much smaller than the applied field. The problem is avoided if the specimen used has a shape with no demagnetizing field, such as a toroid. Fig. 3.15 illustrates this case, in which the specimen is effectively the core of a transformer. By Faraday's law of induction, the electromotive force (EMF) induced in the secondary is equal to the rate of change of flux through it. The flux is proportional to the magnetic induction, B, of the specimen. However, the applied field is proportional to the current in the primary, not to the rate of change of the current. Hence, in order to measure B, the current flowing through the secondary must be integrated. A disadvantage of this method is that very large fields cannot be applied. Its use is therefore confined to soft magnetic materials, which can be magnetized in fairly small fields. This makes the use of toroidal specimens both convenient (because only small fields need to be applied), and necessary (because the large magnetization would otherwise lead to large demagnetizing fields).

Another way to make use of induced currents is to place the specimen inside a set of search coils situated in the magnetizing field. The two most common configurations are shown in Figs. 3.16 and 3.17. In both cases, the specimen is made to move along the line AB. If the specimen has a magnetic moment, its movement induces EMFs of opposite sign in the coils S_1 and S_2 (Fig. 3.16), or in $S_1 + S_2$ and $S_3 + S_4$ (Fig. 3.17). Since these coils are connected in series opposition, the signals are added. This method of connecting the search coils also ensures that signals due to changes in the magnetizing field are cancelled out. The signal produced in the coils is thus proportional to the magnetic moment of the specimen.

71

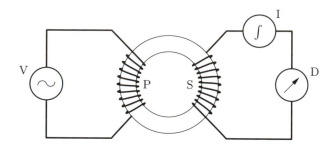

FIG. 3.15   Induction method of measuring the magnetization of a toroidal specimen. (P = primary, S = secondary, I = integrator, V = source of
current in P, D = detector.)

However, the EMF induced depends also on the velocity of the specimen.
To get a signal that depends only on the magnetization and not on the
velocity, the current from the coils can be integrated electronically. The
most accurate method based on this type of apparatus is to vibrate the
sample at a fixed frequency, producing an alternating current through the
coils. This is the principle of the *vibrating sample magnetometer*. All these
search coil methods are suitable for materials in which the demagnetizing
fields are not too significant. They are therefore used mainly for permanent
magnet materials and for weakly magnetic materials.

Magnetization can also be determined by measuring the force exerted
on a specimen by a magnetic field. When a magnetic dipole of moment m
is placed in a magnetic field H, the energy changes by an amount

$$E = -\tfrac{1}{2}\mu_0 mH, \tag{3.27}$$

or

$$E = -\tfrac{1}{2}\mu_0 VMH, \tag{3.28}$$

where M is the magnetization and V the volume of the specimen. If the
energy varies along the coordinate x, then the specimen experiences a force
F, equal to minus the rate of change of E along x:

$$F = -\frac{dE}{dx} = \tfrac{1}{2}\mu_0 V \frac{d(MH)}{dx}. \tag{3.29}$$

For a diamagnetic or paramagnetic material, M varies with H according
to eq. (1.9), so we have

$$F = \tfrac{1}{2}\mu_0 V\chi \frac{d(H^2)}{dx} = \mu_0 V\chi H \frac{dH}{dx}. \tag{3.30}$$

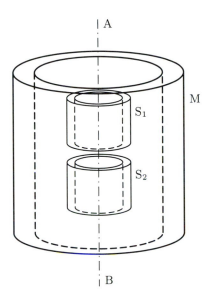

FIG. 3.16   Search coil method of measuring magnetization. The specimen moves
along the axis AB. The coil M produces a field parallel to AB. S_1 and
S_2 are search coils connected in series opposition.

For a permanent magnet in a relatively small field, we can assume M to
be constant, and then eq. (3.29) becomes

$$F = \tfrac{1}{2}\mu_0 V M \frac{dH}{dx}. \tag{3.31}$$

In both cases, it is necessary to have a non-uniform field to produce the
force. As the specimen has a finite volume, it is necessary that the force
acting on different parts of its volume should be the same. Therefore,
in the case of diamagnetic or paramagnetic materials, we must make sure
that $d(H^2)/dx$, or $H(dH/dx)$ are uniform over the volume of the specimen.
In the case of a material with a constant magnetization, dH/dx must be
uniform. In both cases, the field itself may be parallel or perpendicular
to x; the force is in the direction along which the field *varies*, not in the
direction of the field itself.

The non-uniform field may be produced in two different ways. One way
is to use a solenoid or electromagnet to produce initially a uniform field.
A set of coils are then placed in the field to produce an additional non-
uniform field. The second way is to use an electromagnet with non-parallel
polefaces. In either case, the specimen is suspended from a sensitive balance
and the change in its apparent weight is measured when the field is switched
on or off. The method is called the Faraday method.

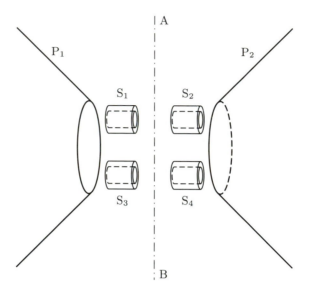

FIG. 3.17 Search coil method of measuring magnetization, using an electromagnet. The specimen moves along the line AB, perpendicular to
the field produced by the magnet (pole pieces P_1, P_2). S_1, S_2, S_3
and S_4 are search coils. S_1 and S_2 are in series, S_3 and S_4 are also in
series, with $S_1 + S_2$ and $S_3 + S_4$ being in series opposition.

An easier version of the force method can be used if the specimen is
available in the form of a long rod. The rod is placed with one end between
the poles of an electromagnet, the other end sufficiently far away from the
poles for the field acting on it to be small (Fig. 3.18). Suppose the field
between the poles is H_1, and the field at the other end of the specimen
is H_2. If the specimen moves vertically upwards by a small distance Δx,
then, from eq. (3.28), the energy at the bottom end changes by

$$\Delta E_1 = \tfrac{1}{2}\mu_0 AMH_1\Delta x$$
$$= \tfrac{1}{2}\mu_0 A\chi H_1^2\Delta x, \tag{3.32}$$

where A is the cross-sectional area of the specimen, since a volume Ax
of the specimen has been replaced by vacuum. At the top end, an equal
volume of vacuum has been replaced by magnetic material, so the energy
changes by

$$\Delta E_2 = \tfrac{1}{2}\mu_0 AMH_2\Delta x$$
$$= \tfrac{1}{2}\mu_0 A\chi H_2^2\Delta x, \tag{3.33}$$

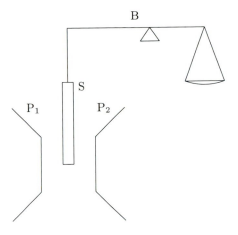

FIG. 3.18   The Gouy method of measuring magnetic susceptibility. The specimen, S, is suspended from a balance, B, between the polepieces P_1
and P_2 of an electromagnet.

The total change of energy is

$$\Delta E = \Delta E_1 + \Delta E_2 = \tfrac{1}{2}\mu_0 A\chi(H_1^2 - H_2^2)\Delta x. \qquad (3.34)$$

The force is given by

$$F = -\frac{\Delta E}{\Delta x} = -\tfrac{1}{2}\mu_0 A\chi(H_1^2 - H_2^2), \qquad (3.35)$$

and, since $H_2 \ll H_1$,

$$F \approx -\tfrac{1}{2}\mu_0 A\chi H_1^2. \qquad (3.36)$$

Again, the specimen is suspended from a sensitive balance, and the change
in its apparent weight due to switching the field on or off is measured. The
force is then directly proportional to the susceptibility, χ. This method
(called the Gouy method) is only suitable for materials whose magnetization is proportional to the field, as it measures susceptibility rather than
magnetization, and it is therefore used mainly for diamagnetic and paramagnetic materials. It is also particularly suitable for measuring the susceptibility of liquids. The liquid is held in a sealed glass tube so that the tube
is exactly half filled, and the tube is suspended so that the top surface of
the liquid is in the maximum field. The net force on the glass is then zero,
even if the glass is more strongly magnetic than the liquid.

A very sensitive instrument based on force methods is the alternating
gradient force magnetometer. The experimental arrangement is similar to
the vibrating sample magnetometer, Fig. 3.17. The sample is suspended in
the centre of the coil system on a very flexible support, such as a thin reed.

In this case however, instead of measuring the induced current in the search coils when the specimen is vibrated, the search coils are supplied with an alternating current. The specimen is made to vibrate by the alternating force produced by this current. If the frequency of the current is equal to the natural frequency of vibration of the specimen and its support, the specimen will resonate, and the amplitude of the vibrations will be large. The amplitude of the vibrations depends on the amplitude of the force, which in turn depends on the magnetic moment of the specimen. The deflection of the specimen is usually measured by a piezoelectric sensor. The high sensitivity of the instrument enables the magnetic moments of very small samples to be measured.

3.3.2 Determination of the Curie temperature

In principle, the Curie temperature of a ferromagnet can be measured in two different ways. Firstly, the susceptibility can be measured as a function of temperature, above the Curie temperature. From eq. (2.8), a plot of χ^1 against T should give a straight line, reaching zero at $T = \theta$ (see also Fig. 2.8). In practice, the line becomes curved near $T = \theta$, and θ has to be found by extrapolation from higher temperatures. The value measured in this way is the paramagnetic Curie temperature (see section 2.3.3). Secondly, the magnetization can be measured as a function of temperature below the Curie temperature. The temperature at which the spontaneous magnetization, M_s, becomes zero is the ferromagnetic Curie temperature. However, in order to measure the spontaneous magnetization, a magnetic field, H, must be applied in order to produce uniform magnetization in the specimen (i.e. drive out the domain walls). This field affects M_s. At low temperatures, the increase of M_s with H is small, but as the Curie temperature is approached, M_s varies more and more strongly with H, as explained in section 2.3.2. The result is that the measured magnetization is greater than M_s, the difference between the two increasing as the temperature increases. At the Curie temperature, where $M_s = 0$, the measured magnetization will not be zero. To determine the Curie temperature accurately, the magnetization should be measured for a number of different values of H as a function of temperature. The results should then be fitted to an appropriate formula from which the measurements can be extrapolated to $H = 0$. There are various ways of doing this. The simplest, though not necessarily the most accurate, is to use the fact that M_s is proportional to $(\theta_C - T)^{1/2}$ near the Curie temperature. A plot of M_s against T is therefore parabolic when $T \approx \theta_C$. Fig. 3.19(a) shows the variation of M_s, and of the measured magnetization, M, with T. The two curves approach each other as T decreases. If we plot M^2 against T, we get a curve that approximates to a straight line (Fig. 3.19(b)), so that θ_C can be found more accurately by extrapolation. More accurate extrapolation methods can be found in the literature (see appendix F).

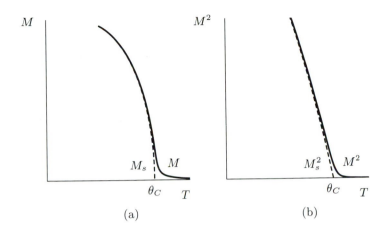

(a)   (b)

FIG. 3.19   Methods of plotting measured magnetization M against temperature T, to determine the Curie temperature. (a) M against T gives a curve approaching a parabola, (b) M_2 against T gives a curve approaching a straight line, as T decreases.

3.3.3   Measurement of anisotropy

One of the ways of measuring magnetocrystalline anisotropy is by measuring the magnetization curves of single crystals with the field applied in different directions. This method only works for soft magnetic materials, which have a small hysteresis. Suppose we have a single crystal of a cubic material in which the easy directions are the $\langle 100 \rangle$ directions (i.e. K_1 is positive in eq. (3.13)). If the field is applied parallel to [100], the crystal is very easily saturated in a small field. The magnetization curve is almost vertical until saturation is reached (see the curve labelled [100] in Fig. 3.7). If the field is applied parallel to [110] (see Fig. 3.7), there is an initial rapid rise of magnetization while the domain walls are swept out of the crystal. Once the crystal has become uniformly magnetized in the easy direction that makes the smallest angle with the field, a further increase of field rotates the magnetization away from the easy direction towards the field. As this rotation is opposed by the anisotropy, the increase of magnetization will be slower. It can be shown that the area between the two curves (a and b) is proportional to the anisotropy constant.

For this measurement to be accurate, demagnetizing fields must be eliminated. This means that the specimen must be in the shape of a 'picture frame' (Fig. 3.20) with the sides being parallel to [100] and [010] for measuring curve (a), and to [110] and [1$\bar{1}$0] for measuring curve (b). The specimen preparation is therefore rather laborious, and this method is rarely used in practice.

The most commonly used method of measuring anisotropy is torque

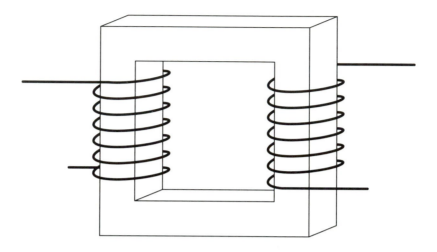

FIG. 3.20   Picture-frame specimen for measuring magnetization curves.

magnetometry. This method still requires a single crystal, but only one specimen is needed which is usually in the shape of a small disc or sphere. The specimen is suspended on a fine wire between the poles of a magnet, in a field large enough to saturate the specimen in the direction of the field. Usually, it is necessary to anchor the specimen with a second torsion wire from a fixed point below the specimen, to prevent it from swinging sideways and sticking to one of the polefaces of the magnet. In the applied field, the specimen tries to rotate so that an easy direction of magnetization becomes parallel to the field. The rotation is opposed by the torsion suspension. In the simplest form of the apparatus, the specimen rotates by a small angle which is proportional to the torque acting on it. This torque can be calculated from the expression for the anisotropy energy (e.g. eq. (3.13) or (3.14), depending on the crystal structure). The energy, E, must be expressed as a function of the angle of rotation about the torsion suspension, ϕ, say. Then the torque is given by $dE/d\phi$. It varies periodically with ϕ. If the magnet is placed on a rotating stand, the torque can be measured as a function of ϕ. The maximum torque is then proportional to the anisotropy constant.

The method can be made more accurate and convenient if instead of measuring the angle by which the specimen turns and then deducing the torque from this angle, we apply a torque in the opposite direction just sufficient to prevent the specimen from rotating, and then measuring this applied torque. The simplest way is to suspend the torsion wire from a rotating dial which is adjusted to keep the specimen in a fixed orientation, and then measuring the amount of adjustment needed as a function of the rotation of the magnet. A better way is to apply the torque electronically. There are a number of simple servo-mechanisms that use a current or a voltage to apply a torque of the correct magnitude automatically. This current or voltage can then be directly displayed on a chart-recorder in the y-direction, while the rotation of the magnet is displayed in the x-direction, so that the torque curve is automatically plotted out. The signals can also be fed into a computer, which can be programmed to calculate the anisotropy constant.

If the material has more than one anisotropy constant, they can all be measured by suspending the specimen in several different orientations. In all these measurements, it is important to avoid extra torques due to demagnetizing effects. In the case of a spherical specimen, there are no such extra torques, but if the specimen is disc-shaped, it must be suspended with the flat surfaces normal to the suspension.

3.3.4 Measurement of magnetostriction

Another important property of magnetic materials is magnetostriction. In section 3.2.5 we qualitatively discussed its effect on the domain structure. When a specimen is taken from a demagnetized state to saturation by applying a magnetic field, its linear dimensions change, reaching a constant value as the specimen reaches saturation. The total fractional change of length is called the linear magnetostriction constant. As explained in section 3.2.5, linear magnetostriction is due to the rearrangement of the domains. Each domain is distorted even when the specimen is demagnetized, but when the magnetization increases, the distortions are aligned in the same direction, producing an overall change of length. There is also a different kind of magnetostriction effect, called volume magnetostriction, which is a change of volume accompanying a change of *spontaneous* magnetization. These changes occur not as a result of applied fields, but as a result of changing the temperature of the specimen. The most rapid change occurs near the Curie temperature.

In this section, we briefly discuss the measurement of linear magnetostriction. Consider first what happens in the case of a polycrystalline specimen. When it is magnetized, it undergoes a fractional change of length, $\delta l/l$, in the direction of magnetization. However, its volume stays approximately constant, and so there must be a fractional change of length of $-\frac{1}{2}\delta l/l$ in directions perpendicular to the magnetization. These changes can most easily be measured by using a strain gauge glued to the specimen. As the specimen is strained, the resistance of the gauge changes in proportion to the strain. The change of resistance can be measured by having the gauge as one arm of a Wheatstone bridge, the other three arms being standard resistances of similar value to that of the strain gauge. However, this method will be found very inaccurate in practice, because the strain is very small, usually about 10^{-5} or less. There are a number of sources of error that must be eliminated if such small strains are to be measured. An important source of error is thermal expansion of the specimen, which can cause strains comparable with the magnetostriction even for small changes of temperature. Another is the fact that in zero field, the specimen is not necessarily demagnetized. A third one is that the resistance of the gauge may be affected by the applied magnetic field, a phenomenon called *magnetoresistance*. This last effect is usually small for metal wire strain gauges except at very low temperatures. However, metal gauges are not very sensitive, and as we need to measure very small strains, it is better to use a semiconductor gauge, which is more sensitive. Semiconductor gauges have a larger magnetoresistance, so it might seem that they would create as many problems as they solve. Fortunately, it is quite easy to eliminate all these sources of error almost completely. Instead of using one strain gauge and three standard resistances to make up a bridge circuit, four strain gauges are used. The specimen is in the shape of a disc, and two of the gauges are glued to one of the flat faces at right angles to each other. (Pairs of gauges at right angles to each other are available as single units.) The other two gauges are glued to a disc of similar size made of a non-magnetic material, also at right angles to each other. The two discs are then mounted coaxially, with the surfaces having the strain gauges being on the inside, facing each other and placed close together but not touching, and the assembly is placed between the poles of a magnet. The two discs should be oriented so that the gauges on one disc are parallel to those on the other. The bridge circuit is connected so that the two gauges on the magnetic specimen form one arm of the bridge. The magnet is placed on a rotating stand, and the changes of resistance are measured as the magnet is rotated. The field is kept constant at a value large enough to saturate the specimen. We are thus effectively measuring the difference between the strains in two perpendicular directions, which varies between $\pm\frac{3}{2}\delta l/l$ as the field rotates through $90°$. This method eliminates all three main sources of error. Effects due to thermal expansion of the specimen are cancelled out by having two gauges on it, and similarly, the two gauges on the non-magnetic specimen cancel out effects due to its thermal expansion. Magnetoresistance effects are cancelled out by having two pairs of parallel gauges in the field. Errors due to the specimen not being completely demagnetized in zero field are eliminated by varying only the direction of the field, not its magnitude.

Magnetostriction effects in single crystals are more complicated. When the specimen is magnetized, the maximum change in its length may occur in a different direction from the magnetization, and the magnitude of the maximum strain may vary with the direction of the magnetization. To characterise the magnetostriction of a single crystal, it is therefore necessary to make measurements with the gauges oriented in several different directions relative to the specimen. However, the principle of using four gauges to eliminate errors still applies.

3.3.5   Observing magnetic domains

If magnetic domains were visible on the surface of a specimen under an optical microscope, their existence would have been discovered long before the
Weiss theory was formulated. But although there are quite a large number
of methods of making domains visible, using a variety of instruments, the
straightforward optical microscopy method does not work, because there is
no reason why the direction of magnetization should affect the appearance
of the specimen surface. There are however two ways in which domains
can be made visible under an optical microscope. Two different techniques
using optical microscopy are described in sections 3.3.5.1 and 3.3.5.2. The
uses of transmission electron microscopy and scanning electron microscopy
for domain observations are discussed in sections 3.3.5.3 and 3.3.5.4 respectively. A special technique in which the electron spin is used for domain
imaging is described in section 3.3.5.5. Sections 3.3.5.6 and 3.3.5.7 discuss
the uses of magnetic force microscopy and X-ray topography respectively,
and section 3.3.5.8 contains some general remarks about observing domain
wall motion.

3.3.5.1   The Bitter technique

The first method, the earliest to be developed, makes use of 'magnetic
liquids', which are colloidal suspensions of very small magnetic particles,
usually of a ferrite such as Fe_3O_4, in a carrier liquid, which may be water
or an organic solvent. In a non-uniform magnetic field, the particles move
towards the positions where the magnetic field is strongest. Near the surface of a ferromagnet, stray fields are present, which are usually strongest
along the lines where domain walls meet the surface. When a drop of magnetic liquid is placed on the surface, the particles tend to delineate the
domain walls, making the domain pattern observable under a microscope.
When the method was first used, in the early 1930s (notably by Bitter,
after whom the method is named), the importance of correct preparation
of the specimen surface was not realised. Mechanically polished surfaces
of metals are heavily strained, and the strains produce an irregular 'maze'
pattern of domains, which is quite unlike the domain structure in the interior of the specimen. In order to eliminate the strains, the surface must
usually be electropolished. In earlier days (until about the early 1970s),
the magnetic liquid had to be prepared by the experimenter, but nowadays
a variety of suitable liquids with different ranges of particle sizes are available commercially. Fig. 3.21 shows an example of magnetic domains on the
basal surface of a hexagonal ferrite crystal image by the Bitter technique.
The specimen is Co_2X ($Co_2Ba_2Fe_{28}O_{46}$, see section 2.4.4.3). There are six
easy directions parallel to the surface. The dark lines are caused by the
agglomeration of magnetic particles along the intersections of the domain
walls with the specimen surface.

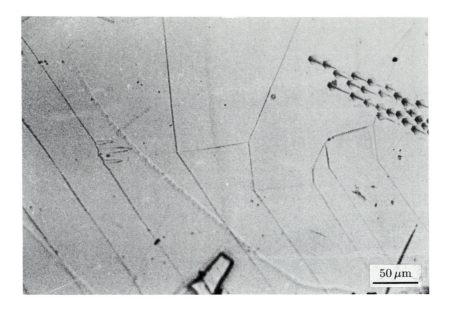

50 μm

FIG. 3.21　Magnetic domains on the basal surface of a hexagonal ferrite specimen, Co_2X, revealed by the Bitter technique. (Courtesy Dr G. A. Jones.)

3.3.5.2　*Polarised light techniques*

The second optical microscopy technique makes use of polarised light. When linearly polarised light is either reflected from a magnetized surface, or transmitted through a magnetized medium, the plane of polarisation is rotated. The angle of rotation depends on the component of magnetization in the direction of the light beam. If the light is reflected from, or transmitted through, two domains magnetized in opposite directions, the plane of polarisation rotates in the opposite sense in the two domains. Using an analyser, it can be set to extinguish the light from one of the domains while allowing some light from the other to pass through. The images of the domains therefore have different intensities. As the angles of rotation are small, the correct setting of the analyser is usually very near, but not exactly at, the crossed position. The effect of the magnetization on the transmitted and reflected light is called the Faraday and Kerr effect respectively.

As the Faraday effect is obviously only suitable for transparent materials, it is restricted to very thin metal films or to fairly thin slices of oxide magnetic materials. The most suitable materials to study using the Faraday effect are garnets, which are good electrical insulators, and are therefore fairly transparent. As the effect depends on the component of magnetization in the direction of the light beam, the method works best on materials in which the domains are magnetized normal to the surface. If the magnetization is parallel to the surface, oblique incidence must be used. An example of magnetic domains in a thin garnet film revealed by the Faraday effect is shown in Fig. 3.22. The method is particularly important in the study of bubble domain devices (see section 4.3.4.1).

Similar considerations apply to the Kerr effect. Normal incidence can only by used for materials in which the magnetization has a component normal to the surface. As the demagnetizing energy favours the magnetization to be parallel to the surface, this configuration (called the polar Kerr effect) can only be used for materials with a strong enough anisotropy to overcome the demagnetizing effect. An example of domains on a basal surface (i.e. parallel to the (0001) plane) of a cobalt single crystal plate imaged by the polar Kerr effect is shown in Fig. 3.23. The specimen is wedge-shaped, with thickness varying from about $15\,\mu$m at the bottom of the picture to about $25\,\mu$m at the top. With increasing thickness the domains become wider in the interior of the specimen and undulating in a more complex way at the surfaces.

In cases where the magnetization is parallel to the surface, the longitudinal Kerr effect can be used, in which the illumination is oblique and the plane of incidence is parallel to the magnetization. Fig. 3.24 shows magnetic domains in an iron-Permalloy multilayer film of total thickness 250 nm, imaged by the longitudinal Kerr effect. This material, which is made up of very thin layers of iron alternating with very thin layers of Permalloy, has magnetic properties well suited to applications as read/write heads in magnetic recording. The two images are of the same area, but with different planes of incidence of the light. The domain magnetization directions are indicated in the upper image. The planes of incidence are chosen so that the four magnetization directions give rise to four different intensity levels in each image. (The intensity depends on the magnitude of the magnetization component parallel to the direction marked as the 'magnetooptic sensitivity' in each image.) Note that the films have an easy axis of magnetization in the plane of the films, perpendicular to their long dimension. The central domains are magnetized alternately 'up' and 'down' in the easy directions, and the triangular closure domains are formed along the edge in order to close the flux within the film (i.e. to decrease the magnetostatic energy). In a few places, extra contrast is visible along domain walls.

3.3.5.3   *Transmission electron microscopy*

There is an important advantage in using electrons rather than light for imaging magnetic domains. As electrons carry electric charges, they experience a force when they move through a magnetic field, or more correctly, a region in which a magnetic induction B is present. The force F has a

83

(a) (b)

(c) (d)

(e) (f)

100 µm

FIG. 3.22 Magnetic domains in a ferrimagnetic garnet specimen, imaged by the Faraday effect. A magnetic field was applied perpendicular to the plane of the specimen in (a) and (b). It was removed in (c) and (d), and applied in the opposite direction in (e) and (f). The analyser was rotated clockwise from the crossed position in (a), (c) and (e), and anticlockwise in (b), (d) and (f).

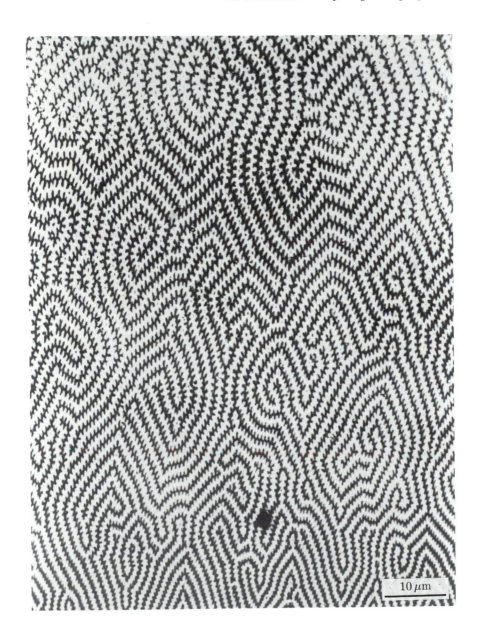

FIG. 3.23 Magnetic domains on a basal surface of a cobalt single crystal plate, imaged by the polar Kerr effect using an oil immersion microscope. The specimen is wedge-shaped, with thickness varying from about 15 μm at the bottom of the picture to about 25 μm at the top. (Courtesy A. Hubert.)

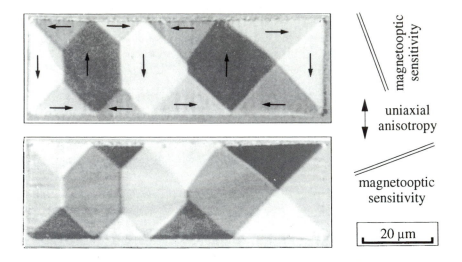

magnetooptic
sensitivity

↑
↓ uniaxial
anisotropy

magnetooptic
sensitivity

20 μm

FIG. 3.24 Magnetic domains in an iron-Permalloy multilayer film of total thickness 250 nm, imaged by the longitudinal Kerr effect. The two images are of the same area, but with different planes of incidence of the light, as indicated by the lines labelled 'magnetooptic sensitivity'. The domain magnetization directions are indicated in the upper image. (Courtesy A. Hubert and M. Rührig.)

magnitude

$$F = -|e|vB\sin\theta, \tag{3.37}$$

where e and v are the charge and the velocity of the electron respectively, and θ is the angle between B and the direction of motion. The force is in a perpendicular direction to both v and B, with its sense being given by the right-hand rule (thumb, first finger and second finger along F, B and v respectively). As electrons pass through a magnetic specimen, this force causes them to be deflected by a small angle. (It can be shown that this angle is given by

$$\beta = \frac{e\lambda Bt}{h} \tag{3.38}$$

for electrons travelling a distance t in a region where the magnetic induction has a component B perpendicular to the motion of the electrons; λ is the electron wavelength and h is Planck's constant. When B is parallel to the plane of the specimen and the electrons are travelling perpendicular to the plane, then t is the thickness of the specimen.) This small deflection can be used to image magnetic domains.

In the transmission electron microscope, electrons pass through a specimen, which may be up to a few hundred nanometres thick. For strongly magnetic materials such as iron or cobalt, the deflection of the electrons (as given by eq. (3.38)) may be up to about 10^{-4} radians. However, the deflections are in different directions in different domains. The magnitude of the deflections is so small the in normal bright-field imaging conditions, electrons can reach the image irrespective of which domain they have passed through, so that the image shows no evidence of the domain structure. However, by altering the imaging conditions slightly, the presence of the domains can be revealed, as shown in Fig. 3.25, which shows four different images of the same area of a specimen of NdFeB magnet material (see section 4.2.4.6). There are two methods by which the domains can be imaged in the transmission electron microscope. Firstly, if the focusing of the image is altered, the images of different domains move sideways in different directions depending on the deflection of the electrons in the specimen, which in turn depends on the magnetization directions. Domain walls then appear as either bright or dark lines, depending on whether the images of the two domains on either side move towards or away from each other. Moreover, a wall the is bright in an overfocused image becomes dark in an underfocused image and vice versa. This effect can be used to distinguish domain walls from other features that may be seen in the image. An example of domains imaged by this defocusing technique is shown in Figs 3.25(a) and (b). In these two images, the domain walls are made to appear as bright or dark lines by defocusing the objective lens in opposite directions. The two images are therefore complementary: the walls that are bright in (a) are dark in (b) and vice versa.

A second method that can be used to reveal the domain structure is to displace the objective aperture by a small distance so that it intercepts the electrons that have passed through one set of domains, while still allowing other electrons to pass through. This method produces images in which the domains themselves, rather than the domain walls, are revealed as bright or dark areas. By displacing the aperture in opposite directions from the microscope axis, different sets of domains can be changed from bright to dark. Figs 3.25(c) and (d) show how domains can be imaged by the displaced aperture technique. In these two images, alternate domains are made to appear bright or dark by displacing the aperture in opposite directions. These two images are also complementary: the domains that are bright in (c) are dark in (d) and vice versa.

3.3.5.4 Scanning electron microscopy

Domains can also be imaged by scanning electron microscopy. This method differs from transmission electron microscopy in its ability to image bulk

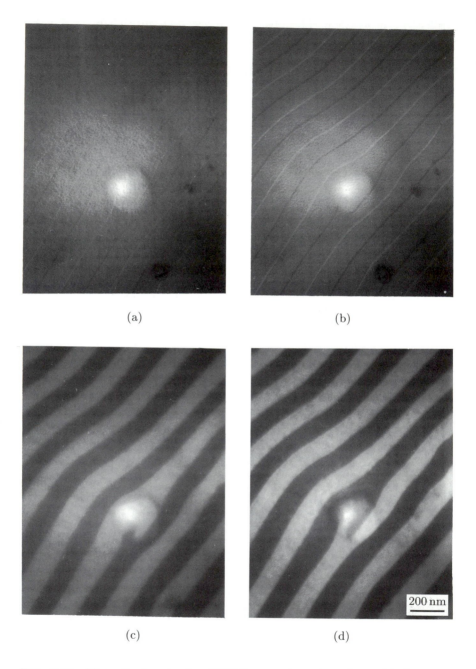

(a) (b)

(c) (d)

200 nm

FIG. 3.25 Magnetic domains in a NdFeB magnet specimen imaged in by transmission electron microscopy. The objective lens was defocused in opposite directions in (a) and (b), and the objective aperture was displaced in opposite directions in (c) and (d).

specimens rather than thin foils. There are a number of different ways in which domains can be imaged by scanning electron microscopy, two of which have been used for detailed domain studies. One of the methods is based on the deflection of secondary electrons as they pass through stray fields near the specimen surface. This method is suitable for strongly anisotropic materials. If the specimen surface is not parallel to an easy direction, the anisotropy keeps the magnetization parallel to easy directions and prevents it from rotating parallel to the surface. There will therefore be free magnetic poles on the surface, which generate stray fields, as illustrated in Figs 3.5(b) and (c). The type of structure illustrated in Fig. 3.8 also gives rise to stray fields above the surface (not shown in the figure). Secondary electrons initially travel away from the surface, but they are deflected by the stray fields either into or out of the plane of the paper in e.g. Fig. 3.5(c). If a secondary electron detector is placed to one side of the specimen chamber (above or below the diagram of the specimen in Fig. 3.5(c)—the microscope axis would be from the top to the bottom of the page, the incident electrons would be travelling downwards on the page), then an image is formed consisting of alternate bright and dark bands, depending on whether the secondary electrons are deflected towards or away from the detector. The brightest and darkest regions are above alternate domain walls, because that is where the horizontal component of the stray fields are largest. Fig. 3.26(a) shows an example of this type of image. The specimen is magnetoplumbite (PbFe$_{12-x}$Al$_x$O$_{19}$, which is an M-type hexagonal ferrite, see section 2.4.4.3). Fig. 3.26(b) was obtained by the same technique, but with the detector in a different position relative to the specimen (relative to Fig. 3.5(c), the detector would be on the right or left of the specimen, in the plane of the paper). In this case, the domains would be invisible if the walls were parallel and straight. The contrast actually seen arises from undulations along the domain walls.

The other method makes use of deflections of the incident electrons after they enter the specimen. Depending on the magnetization direction, the electrons can be deflected towards or away from the surface, and the number of backscattered electrons produced can therefore vary from one domain to another. Contrast between the domains occurs in this case only if the electron are incident obliquely. This method does not require stray fields to be present, and is therefore suitable for studying materials with low anisotropy. However, the image contrast depends on the magnetization, and therefore the method has only been used successfully on strongly magnetic materials. Fig. 3.27 shows a set of four images (a–d) of domains on a (100) surface of iron-3% silicon, a material widely used as transformer cores (see section 4.1.3.1). The electrons were incident obliquely from different directions in the four images, causing the images of different domains to change their intensities in different ways. Note also the additional contrast visible at some domain walls (most obvious where the domains on either side have the same intensity). Fig. 3.27(e) is a schematic diagram of the

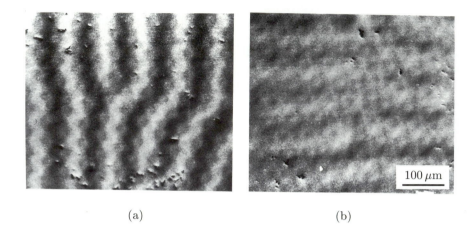

(a)   (b)

FIG. 3.26   Magnetic domains in magnetoplumbite, imaged by scanning electron microscopy. The images were formed using a secondary electron detector placed (a) towards the bottom of the page, and (b) to the right, relative to the specimen.

magnetization configuration, which has been deduced from these images.

3.3.5.5   *Scanning electron microscopy with polarisation analysis*

Electrons emitted from the surface of a magnetic specimen have an unbalanced spin distribution, the majority of electrons having spins antiparallel to the magnetization. A scanning electron microscope fitted with an analyser that can detect the spin direction of the emitted electrons can therefore be used to image the domains. This method can produce images of the domain structure at high resolution. It is again possible to make the detector sensitive to different components of the magnetization. Fig. 3.28 shows two images of the same area of the surface of a silicon-iron specimen. In (a), the intensity depends on the horizontal component of magnetization (M_x, say), whereas in (b), it depends on the vertical component (M_y, say). A schematic diagram of the domain structure and magnetization directions is shown in (c).

3.3.5.6   *Magnetic force microscopy*

The magnetic force microscope is a variant of the various scanning probe techniques that have been developed since the invention of the scanning tunnelling microscope in the 1980s. In the magnetic force microscope, a magnetic probe with a very fine tip is moved parallel to, and very close to

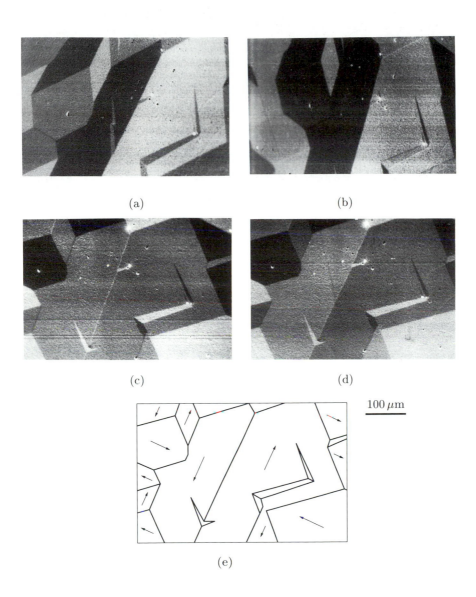

(a)   (b)

(c)   (d)

100 μm

(e)

FIG. 3.27 Magnetic domains on a (100) surface of iron-3% silicon, imaged by scanning electron microscopy. Four different images are shown in (a), (b), (c) and (d), corresponding to four different planes of incidence of the electrons. The plane of incidence in (b), (c) and (d) is rotated by 45°, 90° and 110° respectively relative to (a). A schematic diagram of the magnetization configuration is shown in (e).

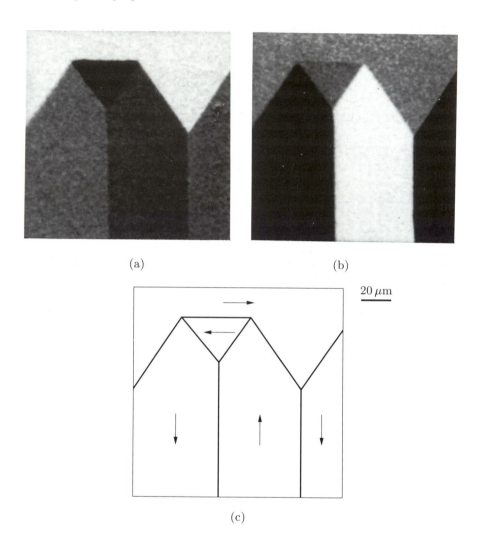

(a)   (b)

20 μm

(c)

FIG. 3.28   Domain structure on the surface of a silicon-iron specimen, imaged in a scanning electron microscope fitted with an electron spin detector. The intensity depends on the horizontal component of the magnetization in (a), and on the vertical component in (b). A schematic diagram of the domain structure and magnetization directions is shown in (c). (Courtesy M. R. Scheinfein.)

the specimen surface, and the force on the needle is measured. As the force depends on the magnetic field at the tip, the method enables the stray field to be mapped, from which the magnetization distribution near the surface can be deduced. Originally, the probe was made by sharpening a fine wire of magnetic material. However, it was found that the probe itself produced a strong stray field which could affect the domain structure being studied. More recent instruments therefore make use of probes with very small magnetic moments, such as a very small magnetic sphere attached to a non-magnetic needle, or a fine cone of non-magnetic material with a thin magnetic film deposited on it. Fig. 3.29 shows an example of a magnetization pattern imaged by magnetic force microscopy. The specimen is a hard disc for a computer, consisting of a thin cobalt alloy film deposited on a non-magnetic surface. The image (a) shows part of a recorded track. The bright and dark stripes correspond to regions magnetized in opposite directions, each stripe representing one 'bit' of information. A schematic diagram of the pattern is shown in (b).

3.3.5.7   X-ray topography

Magnetic domains can also be studied by X-ray topography. Unlike electrons, X-rays are not deflected by the magnetization, but they are very sensitive to small changes in the interatomic spacing or in the orientation of atomic planes. In magnetic materials, such changes occur as a result of magnetostriction. As explained in section 3.2.5, domains are spontaneously strained. If two neighbouring domains are magnetized in opposite directions, they are strained in the same direction, and are therefore not distinguished from each other by X-ray topography. However, if the angle between the magnetization directions is not 180°, the domains are strained in different ways and therefore diffract X-rays differently. An example of magnetic domains in an Fe-3.5%Si specimen imaged by X-ray topography is shown in Fig. 3.30.

For experimental details of X-ray topography, the reader is referred to the literature, but it is useful to compare the advantages and disadvantages of the technique with those of other techniques of domain imaging. As X-rays are fairly penetrating, quite thick specimens can be used. However, X-rays are very sensitive to crystal defects such as dislocations and grain boundaries, and therefore X-ray topography is restricted to single crystals with low dislocation densities. As X-ray topography does not produce magnified images, the photographs must subsequently be enlarged by optical methods, and the resolution is therefore similar to that of optical microscopy techniques, much poorer than that of electron microscopy techniques. X-ray topography generally requires very long exposures, of the order of several hours, if a conventional X-ray source is used. However, very intense sources based on synchrotron particle accelerators can also be used, which reduce exposure times considerably.

A special feature of X-ray topography is that it can be used to image domains in antiferromagnetic materials. It might seem a paradox that domains exist in antiferromagnets, since in the absence of demagnetizing fields, it is not obvious what the 'driving force' is for the formation of do-

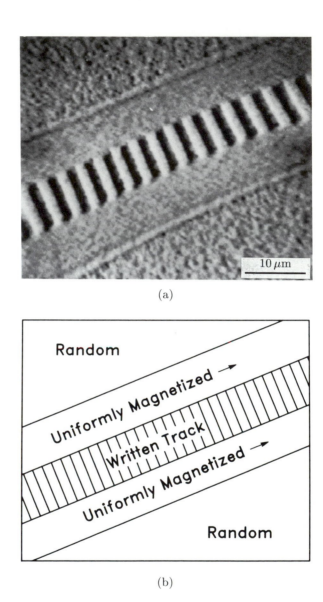

(a)

(b)

FIG. 3.29   (a) Magnetic force microscope image of a magnetization pattern on
a computer hard disc, and (b) a schematic diagram of the pattern.
((a) From D. Rugar, H. J. Mamin, P. Guethner, S. E. Lambert,
J. E. Stern, I. McFadyen and T. Yogi, *J. Appl. Phys.* **68**, 1169 (1990).
© The American Institute of Physics, New York, USA. Reproduced
with permission.)

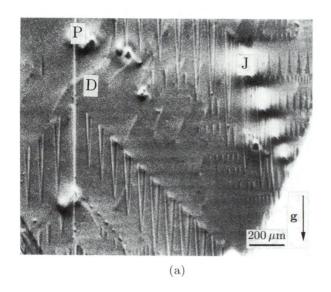

(a)

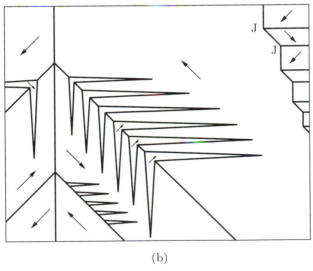

(b)

FIG. 3.30 (a) X-ray topographic image of domains in an Fe-3.5%Si specimen
of thickness $100\,\mu$m and (001) orientation, with (b) an approximate
map of the domain structure. All 180° walls as well as those 90°
walls that are in the vertical direction on the page are invisible, but
triple wall junctions (J) show strong contrast. A precipitate (P) and
a dislocation (D) are also visible. (From J. E. A. Miltat, *Phil. Mag.*
33, 225 (1976). © Taylor and Francis, Basingstoke, UK. Reproduced
with permission.)

mains. However, the three other energy terms, exchange, anisotropy and magnetostriction, all exist in antiferromagnets, and the minimisation of the total energy requires a subdivision into domains. Moreover, antiferromagnetic domain walls can be moved by external stresses, and these movements can be studied by X-ray topography.

3.3.5.8 Moving domain walls

The motion of domain walls can be studied by applying magnetic fields while the specimen is being imaged. In principle, fields can be applied in conjunction with any of the imaging techniques described above. However, some of the techniques are more suitable for studies of domain wall motion than others. In the case of the Bitter technique, the magnetic particles cannot follow domain walls if they move too rapidly, and the pattern becomes fixed once the carrier liquid has evaporated. Electron optical techniques suffer from the disadvantage that electrons are deflected by the applied magnetic fields, causing shifts and distortions in the illumination and the image. These shifts and distortions must be compensated by applying similar fields in the opposite direction away from the specimen. Even so, the magnitude of fields that can be applied is limited, partly by the fact that the compensation is never perfect, and partly by the restricted space available in the specimen chamber for the coils necessary to produce the fields. The techniques that are most easily adapted to dynamic studies are the Faraday effect, Kerr effect and X-ray topography, because images obtained by these techniques are unaffected by the applied fields. However, topography techniques are unsuitable for studies of rapidly changing domain structures because of the long exposure times needed.

3.3.6 Determination of metallurgical parameters by magnetic measurements

As we have mentioned earlier, the metallurgical structure of materials containing one or more magnetic phases affects their magnetic properties, and measurements of these properties can therefore be used to infer structural parameters.

The simplest case is that of an alloy consisting of two phases, both with known magnetization. A measurement of the saturation magnetization of the alloy can then be used to determine the volume fraction. The method is obviously more accurate if the magnetization of the two phases is very different, and is best if one phase is strongly magnetic while the other is non-magnetic. A well-known example occurs in steels in which the body-centred cubic ferrite phase is strongly magnetic while the face-centred cubic austenite phase is non-magnetic. (Note the different use of the word *ferrite* here from the rest of the book, as explained in section 2.4.4.1.) A small amount of ferrite phase may be present in nominally austenitic steels for various reasons, and the amount of this phase may be easily determined by the methods described in section 3.3.1. Similarly, magnetic measurements can be used to determine the amount of austenite in ferritic steels, which may vary for example with tempering.

In other cases, the saturation magnetization as a function of temperature may have to be measured. As explained in section 2.1.2, the saturation magnetization, M_s, of all ferromagnets varies in the same way with temperature, T, if both M_s and T are suitably scaled (Fig. 2.1). In an alloy containing more than one phase, each phase has the same type of dependence of M_s on T, but for each phase, $M_s(0)$ and θ_C are different. The variation of M_s with T for the alloy then shows a more complicated behaviour than that shown in Fig. 2.1. If the volume fraction and composition of the phase is independent of temperature up to the highest Curie temperature of the alloy, then the behaviour of M_s is simply the sum of curves of the type shown in Fig. 2.1 with different scaling parameters. However, if the structure of the alloy is temperature-dependent, then M_s can vary in a more complicated way. In some cases, there may even be an increase in M_s with increasing T in certain temperature ranges. Measurements of M_s as a function of T have been used to determine phase boundaries in some alloys.

Measurements of anisotropy can be used to determine the amount of texture in polycrystalline materials. In the case of rolled sheets, this method measures the average texture, in contrast to X-ray diffraction methods, which measure the texture in a surface layer.

Magnetic measurements, particularly as a function of applied fields, can be used to detect defects in materials. Such methods have been developed into in-situ non-destructive testing of components in engineering structures. Besides measurements of magnetization, permeability or other parameters as a function of locally applied fields, another method that can be used is a variation on the Bitter method of domain observation (section 3.3.5.1) on a larger scale: when a magnetic fluid is applied to the surface, the magnetic particles can delineate cracks formed below the surface.

4

Magnetic materials for practical applications

4.1 Soft magnetic materials

4.1.1 General considerations

ONE of the most important applications of magnetic materials can be described in very general terms as the enhancement of the magnetic effects produced by a current-carrying coil. If a material is to be useful for such applications, it is necessary that it should be easily magnetized. Materials having this property are called soft magnetic materials. The term *soft* refers to their magnetic, not their mechanical properties. However, the conditions in which the material is magnetized can vary widely, and a material that has useful soft properties in some applications, may be quite useless in others. A good soft magnetic material should have a large saturation magnetization, and the magnetization should be large even in relatively small applied fields—in other words, it should have a large permeability. We may need to magnetize the material in different directions, and if it is to have a large permeability in all directions, it should not have a large anisotropy. It should also not have a large magnetostriction, because that can give rise to induced anisotropy. If the material is to be repeatedly magnetized and demagnetized, it should have a narrow hysteresis loop, because each time the loop is traversed, an amount of energy proportional to its area is dissipated. Changing the magnetization leads to eddy-currents being induced in the material. These currents also dissipate energy, and therefore it is advantageous to have a large electrical resistivity. There are other considerations besides magnetic and electrical properties, such as mechanical properties and cost.

There is no material that can score top marks for all these requirements. However, not all the requirements are equally important in all circumstances. The relative importance of the different requirements is determined primarily by the frequency of oscillation of the magnetizing field. At zero frequency, the best material is the one with the largest saturation magnetization. The other properties hardly matter at all. For low-frequency oscillating fields, high permeability as well as large saturation magnetization is needed, and low coercivity also becomes important. As the frequency increases, the importance of high permeability and low coercivity rises relative to that of large saturation magnetization, because the power loss due to hysteresis is proportional to frequency. The power loss due to eddy currents is however proportional to the square of the frequency. It is obvious therefore that at high frequency, the most important requirement is large electrical resistivity.

The plan we will follow in this chapter is to discuss materials suitable for applications at various frequencies, starting from zero.

4.1.2 Electromagnet cores: soft iron and iron-cobalt alloys

The device that makes use of soft magnetic materials in a constant field is the electromagnet. Of the elements, the one most suitable for electromagnet core applications is iron, as it has the largest saturation magnetization, M_s, at room temperature, about $1.72 \times 10^6 \, \mathrm{A \, m^{-1}}$ (saturation magnetic induction, B_s, about 2.16 T). Soft iron is therefore quite a good material for use as electromagnet cores, and is widely used for this application. However, iron usually contains a number of non-metallic interstitial impurities such as carbon, nitrogen, oxygen and sulphur, which can affect its magnetic properties even if they have little effect on the saturation magnetization. In particular, they increase the coercivity and reduce the permeability. The impurity content can be reduced most effectively by annealing in purified hydrogen. This treatment can reduce the coercivity from about $80 \, \mathrm{A \, m^{-1}}$ to about $4 \, \mathrm{A \, m^{-1}}$ and increase the maximum permeability from about 10^4 to 10^5. If a core material with very homogeneous and reproducible properties is needed, purified iron must be used even in d.c. applications. An example of such applications is in electron microscopes, in which the lenses must have high-quality magnetic polepieces. Another advantage of pure iron for this application is that it is sufficiently ductile to be machined to accurate tolerances.

The world record for the largest saturation magnetization at room temperature is held not by any element, but by an alloy of about 35% cobalt, 65% iron ($M_s \approx 1.95 \times 10^6 \, \mathrm{A \, m^{-1}}$, $B_s \approx 2.45 \, \mathrm{T}$). Iron-cobalt alloys however are brittle, because of the formation of an ordered structure. The addition of a small amount of vanadium suppresses the ordering and reduces the brittleness, without affecting the magnetic properties significantly, and the alloy most often used has a composition 49% Fe, 49% Co, 2% V, which has $M_s \approx 1.91 \times 10^6 \, \mathrm{A \, m^{-1}}$, $B_s \approx 2.4 \, \mathrm{T}$. It is expensive because of the high cobalt content, and is used mainly where the reduction of size and weight made possible by the larger saturation magnetization is important, such as in aircraft. The alloy has the trade name *Permendur*, indicating that its permeability remains high even in fairly large magnetizing fields.

This property makes it useful for the vibrating diaphragms in telephone receivers.

As with iron, the magnetic properties of iron-cobalt alloys are adversely affected by non-metallic impurities. Purification can result in a tenfold increase of permeability and reduction of coercivity. The high-purity alloy has the trade name *Supermendur*.

4.1.3 Transformer core materials

4.1.3.1 Iron-silicon alloys

The largest quantity of magnetic material is used in low-frequency a.c. conditions, particularly as cores of mains transformers (50–60 Hz) and motors. Of all magnetic materials, the one that is used in the largest quantity is silicon-iron. At low frequencies, this alloy offers the best combination of good magnetic properties and low cost. Silicon-iron has been of great importance in the development of the large-scale use of electrical power.

The main reason for the improvements achieved by the addition of a few percent silicon to iron is the decrease of both the anisotropy and the magnetostriction, and the increase of the electrical resistivity. Other properties are adversely affected by silicon, but the only property that deteriorates seriously with increasing silicon content is the ductility, which limits the amount of silicon that can usefully be added to iron. There is also a gradual decrease of saturation magnetization and of Curie temperature with increasing silicon concentration, but these decreases are not serious and are more than compensated by the decreases of anisotropy and magnetostriction. The magnetostriction becomes almost zero at 5–6% Si (by weight). This is already a higher concentration than the practical limit set by the increase in brittleness. Unfortunately, the anisotropy does not become zero until an even higher silicon content. Nevertheless, the most important property, the power loss, decreases very significantly even for small silicon concentrations. A decrease by about a factor 2 can be achieved on increasing the silicon content from about 2% to about 3.5%.

As silicon-iron cannot be made completely isotropic, it is advantageous if the field is always applied parallel to an easy direction, say [001], because in this way, a small change of field can cause large displacements of domain walls. It is therefore useful if a crystallographic texture can be developed. A method of achieving favourable texture was developed by N. P. Goss in 1934. His method consists of a combination of hot and cold rolling and annealing, and results in the production of sheets with (110)[001] texture (i.e. surfaces parallel to (110) and the rolling direction parallel to [001]). As this process involves heat treatments at high temperatures, another effect of adding silicon to iron becomes advantageous. Increasing the silicon concentration increases the temperature of the body-centred cubic to face-centred cubic transformation (about 912°C in pure iron) and decreases the temperature of face-centred cubic to body-centred cubic transformation (about 1394°C in pure iron), and in alloys containing more than about 2.5% Si, the transformation is entirely suppressed. The absence of this phase transformation is important for the correct texture to be achieved. Another advantage of the rolling process is that it results in thin sheets: eddy current losses decrease with decreasing thickness, and transformers are therefore usually made of thin laminations.

Transformer cores are usually of a rectangular shape. With a (110)[001] texture, it is not possible to have each arm of the rectangle parallel to an easy direction. In principle, a (100)[001] texture would be better because the arms could then be parallel to [001] and [010] respectively. Methods have been developed for producing such textures, but they have not gained wide acceptance, mainly because of the high production costs. However, because of the enormous commercial importance of silicon-iron, a great deal of research has been done on it recently. The effect of a number of parameters has been studied systematically, such as the nature and amount of impurities, the degree of misorientation of the texture, applied stresses, grain size and thickness, and these studies have lead to a gradual improvement in magnetic properties. A more dramatic improvement came about from the late 1960s onwards with the development, originally in Japan, and later in the United States, of a new high-induction material. There are a number of different manufacturing processes, but they all involve careful control of all variables. Surface coatings play an important role, not only because they insulate neighbouring laminations, but also because the stresses they exert can alter the anisotropy in such a way that the power loss is reduced.

4.1.3.2 Nickel-iron alloys

For applications at higher frequencies, materials with higher permeability are needed, even if their saturation magnetization is not as large. The most commonly used materials in the audio frequency range are nickel-iron alloys. There is a very wide range of useful alloy compositions, including both the binary system and additions of other elements. Alloys based on the nickel-iron system have a wide range of uses, not only as audio-frequency transformer cores.

Nickel-iron has a body-centred cubic structure up to about 30% Ni, and face-centred cubic above. Around 75% Ni, the ordered Ni_3Fe structure forms, but the ordering temperature is near 500°C, so that ordering takes place very slowly. The addition of other elements affects the ordering; some elements, such as chromium, vanadium, copper and molybdenum inhibit it, whereas silicon, germanium and manganese stabilise the ordered structure.

The saturation magnetization decreases from pure iron towards pure nickel, but in addition, there is a very sharp drop at about 30% Ni. The sharp drop is a result of a marked decrease of the Curie temperature near this composition—the magnetization measured at very low temperatures does not exhibit this sharp minimum.

The magnetocrystalline anisotropy is small in the 60%–80% Ni region, but it is very sensitive to ordering, which tends to change K_1 towards negative values. K_1 is zero at about 75% Ni for a disordered alloy, and at about 63% Ni for an ordered alloy. The magnetostriction is small at around 80% Ni.

Historically, the first useful nickel-iron alloys had compositions of around 75%–82% Ni. These alloys (usually called *Permalloys*) have higher permeabilities and lower coercivities, but also lower saturation magnetizations than silicon-iron. Further developments took place in two directions.

Firstly, it was realised that there was an upper limit to the permeability and lower limit to the coercivity because the anisotropy and magnetostriction were not zero at the same Ni/Fe composition ratio. Additions of small amounts of other elements were found to result in reproducible changes of both anisotropy and magnetostriction, and it was possible to develop alloys in which both were zero. Most of these alloys contain 4 or 5% molybdenum, and some also have similar amounts of copper. They have various trade names such as *Mumetal*, *Supermumetal* and *Supermalloy*. They hold the all-time record for highest permeability (maximum permeability up to 3×10^5) and lowest coercivity (down to $0.4\,\mathrm{A\,m^{-1}}$). These alloys are used in high-quality audio-frequency transformers and as magnetic shields.

Secondly, alloys with lower nickel contents were developed. These alloys have larger saturation magnetizations and are cheaper. Although other magnetic properties become less favourable with decreasing nickel concentrations, compositions down to about 36% Ni are sometimes used.

Besides the high-permeability nickel-iron alloys, in which the favourable properties depend on the absence of anisotropy, there are also applications for materials in which anisotropy has been purposely induced. In these alloys, which have cubic crystal structures, induced anisotropy is usually due to *pair ordering*—the preferential alignment of nearest neighbour pairs of similar atoms (Fe-Fe and Ni-Ni) in one direction. Pair ordering leads to a uniaxial anisotropy. If the magnetizing field is applied parallel to the easy direction, the material exhibits a very 'square' hysteresis loop (Fig. 4.1), with high permeability and with ratio of remanence, M_r, to saturation magnetization, M_s, nearly equal to 1 (M_r/M_s is called the *remanence ratio*). If the magnetizing field is applied perpendicular to an easy direction, we tend to get a *skewed* hysteresis loop, with a permeability that is smaller in magnitude, but constant over a wide range of field values, and small values of remanence ratio (see Fig. 4.1). The mechanisms giving rise to these properties have been explained in section 3.2.6. There are two ways to produce pair alignment in nickel-iron alloys. One is by plastic deformation of ordered alloys, and the other is by annealing in a magnetic field. In an ordered alloy, pairs of atoms of particular types are equally distributed in directions related by crystal symmetry. If deformation oc-

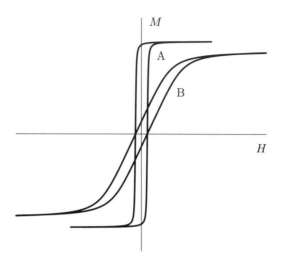

FIG. 4.1   A square (A) and a skewed (B) hysteresis loop.

curs by slip predominantly on one set of parallel slip planes, the nearest neighbour arrangements across these planes are altered, leading to pair ordering. Annealing in a magnetic field (provided the temperature is below the Curie temperature) leads to the alignment of atom pairs that have a smaller energy relative to the field, in the direction of the field.

Anisotropy induced by deformation, in particular by rolling, was first utilised in the 1930s in 50% Ni–50% Fe alloys. Rolling produces an easy direction perpendicular to the rolling direction, and as the material is subsequently magnetized parallel to the rolling direction, the result is a skewed loop. The trade name of this material, descriptive of its constant permeability, is *Isoperm*. Its application is as cores of loading coils.

Magnetic annealing was also introduced in the 1930s, and can be used to produce both square and skewed loop materials. Square loop materials are used in magnetic amplifiers and memory devices.

There are a number of other alloys based on the nickel-iron system that have uses based on their magnetic properties, other than as high-permeability core materials. Alloys with about 30% Ni have a low Curie temperature, as mentioned above, and in the room temperature region, their magnetic properties vary rapidly with temperature. They are used as temperature stabilisers in measuring instruments. Other properties can also vary anomalously near the Curie temperature. Among these are the thermal expansion coefficient and the elastic constants. These anomalies are due to rapid variations of the magnetostriction constants with temperature. In some alloys, the combined effects of lattice thermal expansion and temperature dependence of magnetostriction can compensate each other, producing zero net expansion coefficient over a range of temperatures. These alloys are usually known as *Invars*, and the prototype had a composition of 36% Ni, 64% Fe. The discovery of this alloy around the turn of the century was of great significance. It provided length standards that were more accurate and cheaper than alloys used previously, and also enabled the accuracy of pendulum clocks to be improved. The development of invar alloys earned the Swiss-born physicist C. E. Guillaume the Nobel Prize for physics in 1920. Other alloys have since been developed that have small thermal expansion coefficients.

There are uses for alloys with specific non-zero values of thermal expansion coefficients too. Glass-metal seals, for example, need the thermal expansions of the glass and metal to be the same. The properties of nickel-iron alloys can be tailored to these applications too. In watches and clocks that work with springs, the requirement is for a material whose elastic properties are unaffected by temperature changes. Alloys with such properties also exist in the nickel-iron system, the original alloy having a composition 52% Fe, 36% Ni, 12% Cr. These alloys are called *Elinvars*. The development of all these alloys was also pioneered by Guillaume. Improved Elinvars were subsequently developed.

4.1.3.3 Amorphous alloys

The final group of metallic soft materials we discuss is that of amorphous alloys, also called metallic glasses. They are of special interest, because their development is comparatively recent, and they are still subject to a great deal of exciting research resulting in significant improvements.

The amorphous state in metallic solids is not stable—the crystalline structure is always preferred. However, in a number of alloys, once an amorphous structure is obtained, it has sufficient stability to be retained over long periods. The interest in amorphous magnetic materials originated from the prediction that in the absence of crystal structure, they would have no anisotropy, and would therefore have high permeabilities, low coercivities and low power losses.

Amorphous alloys can be produced in several ways. Soft magnetic materials are usually made by spraying molten alloy onto a moving cold surface such as a rapidly rotating copper wheel, a process called *melt spinning*. This process tends to produce thin, narrow ribbons, but recent developments in production techniques have enabled ribbons several inches wide to be made.

Amorphous alloy soft magnetic materials usually consist of 75% to 85% of iron, cobalt or nickel, or a mixture of two or all three of them, and 15% to 25% of non-magnetic elements, such as boron, carbon, silicon, phosphorus or aluminium, or again, mixtures of them. (The compositions are expressed as atomic percentages.)

The first amorphous alloys made by melt spinning in the early 1970s had disappointingly high coercivities and low permeabilities. The main reasons were compositional inhomogeneities and internal stresses introduced by the rapid cooling—although amorphous materials should have no magnetocrystalline anisotropy, they still have magnetostriction and are thus susceptible to induced anisotropy. From the middle of the 1970s onwards, methods have been developed to improve the magnetic properties after the preparation of the ribbons. The improvements are achieved mainly by annealing, which has to be carried out at sufficiently low temperatures to prevent crystallisation. Annealing tends to homogenise the material, causes a rearrangement of atoms over short distances (a process that has been described as making the material 'even more amorphous') and relieves internal stresses. In some cases, annealing under a tensile stress has been found to give good results.

From the point of view of manufacturing costs, amorphous materials compare favourably with crystalline metallic soft magnetic materials, because once they have been produced, they require no further mechanical treatment such as rolling, and any heat treatment is carried out at low temperatures. As the composition of amorphous alloys can be varied continuously, their magnetic properties cover a wide range. At one end of the range, there are materials with large saturation magnetization, which are potentially competitive with iron and silicon-iron, and at the other, there are materials with extremely low coercivities and high permeabilities, which could take over from nickel-iron alloys in a number of applications.

The largest saturation magnetization occurs in alloys containing the largest amount of iron. However, the maximum values achieved are only about $1.3 \times 10^6\,\mathrm{A\,m^{-1}}$, which is considerably smaller than that of silicon-iron. Their Curie temperatures are also lower. However, their coercivities are about an order of magnitude lower and their permeabilities an order of magnitude higher than that of silicon-iron. Because of the lower saturation magnetization, it is unlikely that amorphous alloys will take over from silicon-iron in large power transformers in the foreseeable future. It is interesting to note that there are differences between the electricity distribution systems used in the United States and Japan, and those used in Europe. In the former, extensive use is made of smaller transformers, whereas in the latter, only large transformers are used. Amorphous materials therefore have a better chance of competing with silicon-iron in the United States and in Japan than in Europe.

For smaller devices, amorphous materials have advantages over crystalline alloys. The alloys with higher magnetization can compete with silicon-iron where higher permeability and lower coercivity are required even at the expense of a somewhat lower magnetization. At the other end of the range, there are amorphous alloys with permeabilities and coercivities only slightly inferior to those of nickel-iron alloys, but with larger magnetization. Amorphous alloys also cover a wide range of other properties such as magnetostriction. The lowest values are found in some iron-cobalt based (cobalt rich) and some iron-nickel based alloys. Another useful property of amorphous alloys is that a uniaxial anisotropy can be induced by annealing in a magnetic field. This treatment allows materials with either square (high remanence ratio) or skewed (low remanence ratio, constant permeability) hysteresis loops to be produced. These materials could compete with the corresponding nickel-iron alloys (see section 4.1.3.2) in similar applications.

4.1.3.4 Cubic ferrites

At high frequencies, metallic core materials are not useful because of large eddy-current losses. The only materials of practical use are insulators. The main types of magnetic insulators were discussed in section 2.4.4. For applications as soft magnetic materials, we need materials with a cubic crystal structure, because other structures generally have larger anisotropies. (There is one exceptional case of hexagonal ferrites that are useful as soft materials, as will be seen later.) There are two types of materials that satisfy these requirements: spinel ferrites and garnets. However, garnets are much more expensive, and are therefore only used for very special applications. By far the largest quantity of insulating soft magnetic materials in use are spinel ferrites.

The requirements are similar to those of metallic soft magnetic materials: large saturation magnetization, high permeability, small coercivity, low power loss and high resistivity—but these properties must be maintained up to as high frequencies as possible.

The manufacture of ferrites is usually carried out in four stages. The first stage is the production of the material with the required chemical composition, using either solid-state reactions in mixtures of oxides, or wet chemical techniques with solutions of salts as the starting materials. The end product of this stage is a powder. The second stage is to compact the powder under pressure, and the third stage is sintering to bind the particles together. At this point, the material has formed into a hard, brittle solid with some porosity. During the fourth stage, the material is machined to the required size and shape. Because of their brittleness, ferrites must be machined by grinding.

Because of the porosity of ferrites, domain walls do not move easily, and magnetization rotation processes play an important role. In order to achieve high permeability, it is important to keep both the anisotropy and magnetostriction small. Of the five pure ferrites, cobalt ferrite is ruled out because it has a large anisotropy, and magnetite (which is effectively 'iron ferrite') because of its low electrical resistivity. The low resistivity is due to the presence of ions of the same element with different valencies (Fe^{2+} and Fe^{3+}), enabling electrons to 'hop' from one to the other. This leaves three ferrites, manganese, nickel and copper, as candidates for practical applications. As described in section 2.4.4.2, the addition of zinc ferrite raises the magnetization, and all ferrites used in practice contain additions of zinc ferrites. Of all the ferrites, manganese-zinc ferrites have the largest magnetization, and by careful control of composition (i.e. the proportion of iron, manganese and zinc), it is possible to achieve very low anisotropy and magnetostriction. In the lower frequency range (up to about 10 MHz), manganese-zinc ferrite is the most widely used material. Its usefulness at higher frequencies is limited by the fact that the compositions at which the anisotropy and magnetostriction are smallest are comparatively rich in iron. This leads to lower resistivity because of the presence of both Fe^{2+} and Fe^{3+} ions. For applications at frequencies above about 10 MHz, nickel-zinc ferrites are usually used.

The permeability of these ferrites is approximately independent of frequency up to a critical frequency, and then decreases exponentially. The power loss has a maximum at about the critical frequency. The critical frequency is the approximate upper limit of usefulness of the ferrite. The properties of nickel-zinc ferrites can be controlled by varying the Zn/Ni concentration ratio. By varying the ratio from about 1.9 to about 0.015, the permeability decreases from about 700 to 10, but the critical frequency increases from about 10 MHz to 100 MHz.

Manganese-zinc ferrites and nickel-zinc ferrites (usually marketed under the trade name *Ferroxcube* of various grades) are used in a wide variety of applications, such as cores for receiving antennas and inductors, wide-band and pulse transformers and cores of recording heads. Television sets and portable radios contain several components made of these ferrites.

4.1.3.5 High-frequency materials

In the region of 100 MHz and above, we get into the realm of more specialised applications, mostly connected with resonance phenomena and microwave devices. We briefly mention two types of materials whose properties make them especially suitable for high-frequency applications. The development of the first is an interesting example of a theoretical prediction being realised in practice. It was explained in section 2.2.1 that when a magnetic field is applied at an angle to a magnetic moment, the moment tries to precess about the field. If this precession could continue unhindered, the angle between the moment and the field would stay constant, and the moment would not align with the field. When a field is applied to a magnetic material, the magnetization changes, which means that the moments do rotate towards the field. This happens because the precession is damped by interactions between the atoms in the material. However, precession does take place, even if only for a short time, and it delays the response of the material to changes of applied field. It was predicted that crystalline anisotropy could be used to prevent precession, thus accelerating changes of magnetization. In a material that had a uniaxial anisotropy but a negative anisotropy constant, the magnetization is free to point in any direction in the plane normal to the symmetry axis, but rotation of the magnetization out of this plane is opposed by the anisotropy. If an oscillating field is applied parallel to the 'easy' plane, the magnetization can rotate within this plane, but cannot precess because of the anisotropy. Materials with such anisotropies can be found among the hexagonal ferrites (an example being Co_2Z, see section 2.4.4.3), and are known under the name of *Ferroxplana*. Their permeabilities are somewhat inferior to the Ferroxcubes at low frequencies, but their critical frequencies are higher, around 1 GHz.

The second group of materials specially suitable for high-frequency applications are the garnets. They have smaller magnetizations than ferrites and are more expensive, but they have one important advantage. As they contain only trivalent ions, they have higher resistivities, and this is especially useful at high frequencies. The higher resistivity causes them to be optically more transparent as well, which makes them suitable for other specialised applications. One such application will be discussed in section 4.3.4.1.

4.2 Hard magnetic materials

4.2.1 General principles of permanent magnets

Our ancestors' awareness of magnetism began when they noticed the unusual properties of lodestone. The attraction of lodestone for pieces of iron and its ability to orient itself relative to the earth are effects we now associate with permanent magnets. Yet, as we have seen in the previous chapter, magnetite, the material of lodestone, is now thought of as a soft rather than a hard magnetic material. (*Hard* magnetic materials are hard to magnetize or to demagnetize, and therefore they make good permanent magnets. The term describes the magnetic, not the mechanical properties.) The distinction between these two categories of materials was arrived at comparatively recently in the history of magnetism. From the modern viewpoint, the only magnetic property equally desirable in both soft and hard materials is a large spontaneous magnetization. (Low cost, reasonable mechanical properties and low density are also desirable, obviously.) All other properties that were stated to be advantageous in soft magnetic materials are either of little relevance (for example, high resistivity, high permeability and low magnetostriction) or disadvantageous (for example, low coercivity and low anisotropy) in permanent magnet materials.

It is fairly easy to realise that in a permanent magnet, a large remanence is even more important than a large saturation magnetization—a permanent magnet must usually operate without an external field helping to maintain its magnetization. However, the requirements are more stringent than this. When a material is magnetized and the magnetizing field is then removed, there is always a field present which tries to demagnetize the material. As we saw in section 3.2.2, this demagnetizing field leads to the formation of a domain structure, resulting in a reduction of the magnetic moment of the specimen. A permanent magnet must be able to resist being demagnetized by its own demagnetizing field (or by an external demagnetizing field, such as is present when two magnets repel each other). A measure of the ability to resist magnetization is the coercivity, and this is why large coercivity is an important requirement in permanent magnets.

In section 3.1, we defined the coercivity as the reverse field needed to reduce the magnetization to zero after saturation (represented by the distance OD in Fig. 3.1; the coercivity is always counted as being positive). This can be regarded as the physicist's (or materials scientist's) definition. The engineer, who designs equipment to make use of magnets, is more interested in magnetic induction, B, than in magnetization, M, because B is what is usually directly measured. The coercivity can therefore also be defined as the reverse field needed to reduce B to zero—the engineer's definition. The two different coercivities are denoted by $_MH_c$ and $_BH_c$ respectively. An important point is that the better the magnet material is, the greater is the difference between $_MH_c$ and $_BH_c$. Let us take the extreme case of a perfect magnet that just cannot be demagnetized. Its remanence is M_r, and when a reverse field is applied, the magnetization remains constant at this value. But from eq. (1.6) we have

$$B = \mu_0(H + M_r),\tag{4.1}$$

so that B becomes zero when

$$H = M_r,\tag{4.2}$$

and the coercivity is

$$_BH_c = M_r.\tag{4.3}$$

Suppose that M remains equal to M_r up to a large reverse field, and then suddenly drops to zero. The field at which this happens is $_MH_c$. Provided that $_MH_c > M_r$, then $_BH_c$ is always equal to M_r, however large $_MH_c$ is.

An important quantity is the product HB, which is equal to the energy per unit volume. If we assume that H is uniform inside the magnet, then the energy of the magnet is $-HB$ times the volume, the negative sign signifying that H is in the opposite direction to B. From eq. (4.1), the energy per unit volume, U, is

$$U = -\mu_0 H(H + M_r).\tag{4.4}$$

Fig. 4.2 shows a plot of U for an ideal magnet. It is seen that U is a parabola, and that it has a maximum value

$$U_{\max} = \tfrac{1}{4}\mu_0 M_r^2.\tag{4.5}$$

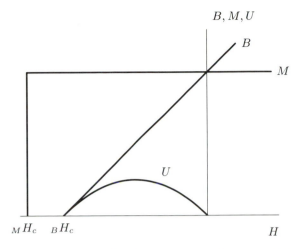

B, M, U

B

M

U

$_M H_c$ $_B H_c$

H

FIG. 4.2   The variation of M, B and U with H, for an ideal magnet.

For a non-ideal material, U_{max} is smaller than this value. In general, it can be taken to be a measure of the quality of the magnet material. We see that for an ideal magnet material, U_{max} depends only on M_r. Of course, if M is constant for positive values of H too, then the remanence is the same as the saturation magnetization, M_s, so that

$$U_{\mathrm{max}} = \tfrac{1}{4}\mu_0 M_s^2. \tag{4.6}$$

The quality of an *ideal* magnet material therefore depends simply on M_s. It is seen that there is an absolute maximum for U_{max} at room temperature, set by the fact that the material with the largest M_s at room temperature is the cobalt-iron alloy discussed in section 4.1.2. It is very unlikely that this maximum will ever be exceeded.

Real materials are non-ideal, i.e. M does not stay constant but varies with H, and so U_{max} is always less than the value given by eq. (4.6) or even eq. (4.5). In practice therefore, it is still worth looking for materials with larger values of $_M H_c$ as well as M_s and M_r.

The maximum energy product, U_{max}, is only an average measure of quality. It can be used to compare two materials without knowing the exact circumstances in which they are being used. It may well happen that a material with a smaller U_{max} will perform more efficiently than one with a large U_{max} in a particular application. To see how this might happen, we plot the part of the hysteresis loop in the second quadrant of the coordinate system, i.e. the quadrant in which $M > 0$, $H < 0$. This is the relevant part of the hysteresis loop when there is no external field applied, and is called the *demagnetizing curve*. Fig. 4.3 shows the demagnetizing curves of two different materials, one with a large remanence and small coercivity (curve A), the other with opposite characteristics (curve B).

111

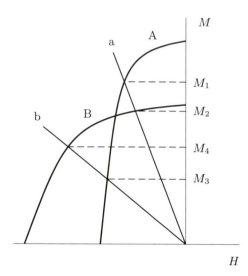

FIG. 4.3   Using the demagnetizing curves to compare different materials.

The demagnetizing field is proportional to M, and is therefore represented by a straight line with negative gradient through the origin. The slope of the line depends only on the shape of the magnet. Suppose we want to compare two magnets of cylindrical shape, magnetized along their axes, with two different values of the ratio of length, l, to diameter, d. Line a is appropriate to the magnet with a larger ratio, line b to that with a smaller ratio. The magnetization is given by the intersection of the straight line with the demagnetizing curve. It is seen that if we need a magnet with a large l/d ratio, we should use material A since $M_1 > M_2$, but if we need a magnet with a small l/d ratio, material B is better since $M_4 > M_3$. This is why the value of $U_{\max}$ is not necessarily the best criterion by which a material for a particular application should be selected.

4.2.2   Ferrous alloy magnets

4.2.2.1   Permanent magnet steels

For a very long time, lodestone was the only known permanent magnet. With the development of steels during the last century came the realisation that some steels were better permanent magnets than lodestone. The best steels appeared to be the ones with fairly high carbon contents, about 1%. These steels were the first artificial permanent magnets, although they were not specifically manufactured for that purpose. The first purpose-made magnet materials were steels with special additions such as tungsten and chromium to improve the magnetic properties, particularly the coercivity.

Further improvements came with the addition of cobalt. These materials have now almost completely disappeared from use, although cobalt steels in particular have the advantage of being more easily machinable to intricate shapes than most of the modern materials. The coercivity in these materials is mainly due to domain walls being impeded in their motion by inclusions of second phases. Typically, these alloys had a remanence, M_r, of about $700 \, \text{kA m}^{-1}$. Coercivities, H_c, ranged from about 4 to $20 \, \text{kA m}^{-1}$ and U_{max} from about 1.6 to over $7 \, \text{kJ m}^{-3}$ from carbon steel to the best cobalt steels.

4.2.2.2   Alnico alloys

From the early 1930s onwards, a completely new class of magnet materials was developed. These materials have the collective name of *Alnico*, after the three elements aluminium, nickel and cobalt, which they contain in addition to iron. The earliest alloys did not actually contain cobalt, and are sometimes known as *Alni* alloys, though sometimes they are included in the group of Alnicos because metallurgically they are similar to the the the cobalt-containing alloys.

A large number of different types of Alnicos have been developed during the last half-century. They differ in composition, additional elements, manufacturing processes and properties. Classification of the alloys is made more difficult by the fact that in different countries, manufacturers use different trade names. In some cases, different names have been given to the same material in different countries, but in others, materials produced in one country do not correspond exactly to any material produced in another. The simplest system of names is probably the one used in the United States, where all alloys are referred to as Alnico followed by a number. This is the system we will follow here. The main types of Alnico are the ones numbered from 1 to 9.

Alnico alloys may be produced by one of two processes, casting or sintering. Either process may be used to produce any of the types of Alnico. Magnets produced by casting usually have somewhat better properties, but sintering is a cheaper process. Sintering is particularly suitable for producing large numbers of small magnets.

In all Alnicos, optimum magnetic properties are produced after the alloy has undergone spinodal decomposition. This results in a two-phase structure, with the phases very finely intermixed. One phase consists predominantly of iron or iron-cobalt, the other of nickel-aluminium. Both phases are body-centred cubic, and are usually called α_1 (Fe-Co rich) and α_2 (Ni-Al rich) respectively. The morphology of the microstructure varies from one type of Alnico to another. In all cases, the iron or iron-cobalt phase is strongly magnetic. It is not certain whether the nickel-aluminium phase is weakly magnetic or non-magnetic; in some alloys at least it is weakly magnetic. In any given specimen, the morphology and composition of the phases can be varied by heat treatment. Some composition changes may even occur reversibly.

In the earliest Alnicos that were developed (Alnicos 1, 2, 3 and 4), optimum magnetic properties were induced by suitable heat treatment, in which the main steps were controlled cooling or quenching from a high temperature followed by tempering at around 600°C to 700°C. Some of these alloys did not contain cobalt, but when the beneficial effects of cobalt and of small amounts of copper were recognised, these two elements were also added. Some alloys also contain titanium, which generally increases the coercivity but decreases the remanence of Alnicos. The heat treatment results in the spinodal decomposition into α_1 and α_2 phases. In some alloys, the α_1 phase becomes elongated in $\langle 100 \rangle$ directions, but as there is no preference for any of the crystallographically equivalent directions, the materials are isotropic, i.e. their magnetic properties are independent of the direction in which the field is applied. The best of these Alnicos have $M_r \approx 500\,\mathrm{kA\,m^{-1}}$, $H_c \approx 70\,\mathrm{kA\,m^{-1}}$, $U_{\mathrm{max}} \approx 15\,\mathrm{kJ\,m^{-3}}$.

Further improvements in magnetic properties were achieved in three different ways: firstly by magnetic annealing (i.e. annealing in a magnetic field), secondly by further variations of composition and thirdly by directional grain growth.

Most modern Alnicos (5, 6, 8 and 9) contain 7–8.5% Al, 12–16% Ni and 2–4% Cu. Cobalt contents vary from 23% to 40% and titanium from 0 to 8%. Alnicos 5, 6, 8 and 9 are treated by magnetic annealing, the effect of which is to single out one of three $\langle 100 \rangle$ directions for the elongation of the α_1 particles. The magnetic properties improve in the direction of the field applied during annealing, and all these materials are therefore anisotropic. Alnico 5 is a random-grain material with no significant amount of titanium, with $M_r \approx 1000\,\mathrm{kA\,m^{-1}}$, $H_c \approx 50\,\mathrm{kA\,m^{-1}}$, $U_{\mathrm{max}} \approx 40\,\mathrm{kJ\,m^{-3}}$. Alnico 6 contains some titanium and has a larger coercivity but a smaller energy product. Alnico 8 has an energy product similar to Alnico 5, with larger coercivity ($\approx 130\,\mathrm{kA\,m^{-1}}$) but smaller remanence, achieved by increasing both the cobalt and the titanium content. Directional grain growth improves the properties of both alloys having compositions corresponding to that of Alnico 5 and Alnico 8 respectively. The latter material (i.e. with composition corresponding to Alnico 8 but directionally grown), is called Alnico 9, which can have U_{max} as large as $75\,\mathrm{kJ\,m^{-3}}$. The required texture is one with each grain having one of its cube edge directions parallel to that of other grains. The magnetic anneal must then be carried out with the field parallel to this direction.

Alnico alloys have reached a stage where little improvement of their properties has been achieved in the last two or three decades. Nevertheless, research is still being carried out on them, as their properties are still not fully understood. In order to understand the magnetic properties, a detailed knowledge of the microstructure is necessary. Although a great deal of information has been obtained by X-ray diffraction, electron microscopy and other techniques, important new results are still being obtained by field ion microscopy and atom probe analysis. With these techniques, the composition of the α_1 and α_2 phases has been measured accurately, and the morphology of the microstructure has been revealed in great detail. In Alnico 5, both the α_1 and α_2 are found to be interconnected—the structure does not consist of elongated α_1 particles in an α_2 matrix. In Alnico 8, the α_1 phase occurs in a much more regular arrangement. It had previously been thought that changes of magnetization in Alnico took place by magnetization rotation in the α_1 phase, of the type discussed in section 3.2.8. However, it now seems that domain wall motion plays a more important role in Alnico than had been thought previously. After more than half a century of research, Alnicos are still presenting interesting problems.

4.2.2.3 *Iron-cobalt-chromium alloys*

Another alloy system with certain similarities to the Alnicos is the iron-cobalt-chromium system. It also consists of two phases, formed by spinodal decomposition. Its main advantages over Alnicos are that useful properties can be obtained in alloys with a lower cobalt content, and that during its manufacture, it passes through a ductile state in which it can be machined. During the 1970s, when the price of cobalt was relatively high, there was increased interest in these alloys, but in the end, they have not succeeded in replacing the Alnicos. A disadvantage of the iron-cobalt-chromium system is that it requires rapid quenching at one stage in its manufacture, which limits the size of magnets that can be produced.

4.2.3 General principles of powder magnets

Alnico magnets were discovered more or less by accident—a new type of steel turned out to have useful magnetic properties. Most of the other hard magnetic materials were developed as a result of deliberate planning. The chief inspiration for new developments was the publication of the theory of magnetization rotation by Stoner and Wohlfarth in 1948 (see section 3.2.8). The theory predicted a coercivity of $2K/\mu_0 M_s$ for a material of uniaxial anisotropy with anisotropy constant K and saturation magnetization M_s— if the magnetization rotates coherently and the field is applied in the direction of the easy axis. This provided a recipe for permanent magnets: any collection of anisotropic particles that are too small to contain a domain wall should have a large coercivity. The anisotropy could be due to the crystal structure of the material, or to the shape of the particles. This theory was the starting point for the development of magnets based on a wide variety of materials, as well as of materials for magnetic recording (see section 4.3.3). Yet none of the hard magnetic materials so far developed has a coercivity anywhere near as large as $2K/\mu_0 M_s$. The reason is that in practice, the magnetization always seems to find an easier way to reverse than by coherent rotation. There are a number of possible reasons for this.

Firstly, if the particles are too large, they are more likely to be non-uniformly magnetized. Even if a particle is not large enough to contain domain walls in zero field, its magnetization may pass through non-uniform states while the field is changing. On the other hand, if the particles are too small, the magnetization can rotate spontaneously because thermal agitation can overcome the effects of anisotropy. (This phenomenon is called *superparamagnetism.* Superparamagnetic materials differ from 'ordinary' paramagnetic ones in having quite large susceptibilities and being able to be saturated in moderate fields, but they also differ from ferromagnets in having no hysteresis, i.e. they have zero coercivity. Their properties are described by eqs (2.4), (2.5), (2.6) and (2.7) but with m much larger than the magnetic moment of a single atom, indicating that groups of magnetic moments remain parallel to each other while their orientations fluctuate randomly.) The coercivity can therefore only approach $2K/\mu_0 M_s$ for a fairly narrow range of particle sizes, which may be difficult to achieve in practice.

Secondly, any crystal defects can cause the magnetization to be non-uniform. Magnetization reversal can nucleate at defects.

Thirdly, the magnetization will only be uniform if the field acting on each particle is uniform. The demagnetizing field is only uniform for a particle in the shape of an ellipsoid. For different shapes, the non-uniform demagnetizing field will tend to induce non-uniform magnetization distributions. Even for an exact ellipsoid, only the field inside the ellipsoid is uniform. Outside, there is a non-uniform stray field. When the particles are compacted together, each particle experiences the non-uniform field of its neighbours.

Although the coercivities of real materials are much smaller than the theoretical upper limit, the knowledge that such a limit exists, which may be very high, has indicated the way to find good magnet materials. Materials that were developed on the basis of this theory fall into two categories according to the origin of their anisotropies. Different criteria for the optimization of magnetic properties apply to materials in the two categories.

If the anisotropy is due to the crystal structure, then the coercivity is inversely proportional to M_s, whereas the remanence is approximately proportional to M_s. Therefore $U_{\max}$, which is of the order of $H_c \times M_r$, is approximately proportional to K. The search for better permanent magnets has therefore meant a search for materials with larger and larger anisotropies. However, as we saw at the beginning of this chapter, $U_{\max}$ does not increase indefinitely with increasing H_c, and hence with increasing K. Once a material with very large K has been found, $U_{\max}$ can only be increased further by increasing M_r.

On the other hand, if the anisotropy is due to the shape of the particles, then it is proportional to M_s^2. Therefore H_c is proportional to M_s, and

$U_{\max}$ to M_s^2. In this case, the best material is simply the one with the largest M_s. However, if the particles are packed too tightly, they can be influenced by each other's stray fields, and in practice, there is an optimum value of the packing density for useful permanent magnet properties.

4.2.4 Powder magnets and other modern magnet materials

4.2.4.1 Hexagonal ferrites

The first permanent magnet materials developed on the basis of the Stoner-Wohlfarth theory were the hexagonal ferrites. Although it had been known earlier that mixed oxides of iron (Fe_2O_3) and barium, strontium or lead were magnetic, it was in the early 1950s that the materials with the formula $(MO)(Fe_2O_3)_6$, where M is Ba, Sr, Pb, or a mixture of them, were found to have a strong uniaxial anisotropy. These materials, particularly barium ferrite, formed the basis of the first powder magnets. Today, ferrite magnets are the most important of the hard magnetic materials from the point of view of amount manufactured. They have a number of advantages: the raw materials are relatively cheap, manufacturing processes are fairly simple and require much less care than those of metallic materials, they are chemically stable, and have low density. They have fairly large coercivities, larger than those of the Alnicos. On the debit side, their saturation magnetizations and remanences are small compared with other hard magnetic materials, and although their Curie temperatures are high enough, their magnetization decreases rather rapidly with increasing temperature even around room temperature.

The main steps in the manufacture of barium ferrite magnets are (1) production of the compound with the required composition, (2) breaking up into powder, (3) pressing, (4) sintering, (5) forming into the required shapes, usually by grinding, and (6) magnetizing. This is a simplified description; some of these steps are themselves subdivided into several different processes. Step (2) tends to produce plates with surfaces parallel to the basal plane. Pressing, step (3), therefore produces some alignment with the [0001]-axes of particles tending to be parallel to each other. The alignment is greatly enhanced if pressing is carried out in a magnetic field. Pressing is usually done in a die, and the die can be placed inside a coil which produces the field. The die is designed with some magnetic and some non-magnetic components distributed in such a way that a uniform field is concentrated at the powder being pressed. Magnets in which the particles are aligned are called anisotropic, and have superior properties to isotropic, i.e. non-aligned materials, though there are some applications in which the latter are acceptable. Typical magnetic properties are $M_r \approx 150\,kA\,m^{-1}$, $H_c \approx 210\,kA\,m^{-1}$ and $U_{\max} \approx 6.5\,kJ\,m^{-3}$ for isotropic materials and $M_r \approx$ 250 to $310\,kA\,m^{-1}$, $H_c \approx 135$ to $280\,kA\,m^{-1}$ and $U_{\max} \approx 20$ to $25\,kJ\,m^{-3}$ for anisotropic materials. (For the latter, larger M_r usually goes with smaller H_c and vice versa. The coercivity values are $_M H_c$; values of $_B H_c$ of isotropic materials are lower than those of anisotropic materials.)

Ferrite magnets can also be manufactured in the form of composite materials, embedded in a non-magnetic carrier. The carrier can be a resin, giving a hard composite, or one of a number of plastics or even rubber, making the magnet flexible. In either case, the material can be made isotropic or anisotropic. The magnetic properties of composite magnets are somewhat inferior, but their mechanical properties superior to those of sintered ferrites.

4.2.4.2   Elongated single domain particle magnets

In the late 1950s, a process was developed for producing magnets from elongated small particles of materials that are magnetically soft in bulk, such as iron and iron-cobalt alloys. The process consists of an electrolytic method of precipitating the powder in mercury and subsequently coarsening it by a low-temperature anneal. As we have seen above, the most important requirement for a magnet consisting of particles whose anisotropy is due to their shape is a large saturation magnetization. The upper limit of the anisotropy for an elongated single-domain particle is $\frac{1}{2}\mu_0 M_s^2$, and therefore the upper limit of the coercivity is M_s. This value is however never achieved in practice. The largest coercivity for an iron-cobalt magnet is about $70\,\mathrm{kA\,m^{-1}}$, and the largest $U_{\max}$ is about $25\,\mathrm{kJ\,m^{-3}}$. These magnets are called ESD ('elongated single domain'). For iron magnets, for which the raw materials are cheap, the properties are poorer. Attempts have also been made to produce magnets from cobalt particles. It was hoped that ESD cobalt magnets would have both shape anisotropy and magnetocrystalline anisotropy—hexagonal cobalt has an anisotropy constant somewhat larger than that of barium ferrite. Unfortunately, small particles of cobalt have always turned out to have a face-centred cubic structure, and no one has succeeded in producing hexagonal cobalt powder. Iron and iron-cobalt are the only materials from which ESD magnets have been produced. In most other hard magnetic materials apart from the Alnicos and iron-cobalt-chromium, it is the magnetocrystalline anisotropy that is exploited.

4.2.4.3   Platinum-cobalt alloys

In the late 1950s, a new material emerged, which had the best permanent magnet properties achieved up to that time. This material was an alloy of platinum and cobalt in equiatomic proportions. Since it contains much more platinum than cobalt by weight, it is very expensive, and it only ever found applications in devices requiring very small magnets. Platinum-cobalt can undergo an order-disorder transformation when suitably heat treated. The disordered form is face-centred cubic, but the ordered form is tetragonal, with the same structure as copper-gold, CuAu. The tetragonal form has a large uniaxial anisotropy, and powders made from this material can have very good magnetic properties, up to $M_r \approx 510\,\mathrm{kA\,m^{-1}}$, $H_c \approx 400\,\mathrm{kA\,m^{-1}}$, $U_{\mathrm{max}} \approx 76\,\mathrm{kJ\,m^{-3}}$. Good magnetic properties can also be achieved in bulk by suitable heat treatment. When the material is annealed below the ordering temperature, its coercivity passes through a maximum, then decreases again. The maximum is thought to correspond to a partially ordered state, with tetragonal inclusions in a cubic matrix, and the coercivity is thought to be due to domain wall pinning by the tetragonal inclusions. The magnetic properties of platinum-cobalt have now been exceeded by materials that are cheaper as well, so the demand for it has almost completely disappeared. It has however one advantage over the newer materials: it is chemically much more stable.

4.2.4.4 Rare earth-cobalt alloys

The materials that took over from platinum-cobalt alloys in the early 1970s in nearly all applications requiring high-quality magnets were the rare earth-cobalt alloys. Their development was the result of a large amount of fundamental research. The properties of a large number of alloy systems containing a rare earth element and a transition metal were studied systematically. The subject was extremely large: not only are there over a dozen rare earth and related elements to be combined with each of the three magnetic transition metals, but within each binary system, there are a number of compositions where stable intermetallic compounds exist. Within the transition metals, cobalt turned out to be the most promising because its alloys were generally more anisotropic. Of the rare earths, the lighter ones, with atomic numbers ranging from 58 (cerium) to 62 (samarium) were more favourable because their alloys with cobalt had larger saturation magnetizations than those of the others. Of these alloys, some turned out to have negative anisotropies (magnetization parallel to the basal plane being favoured), making them unsuitable for permanent magnets. The number of possible alloys was therefore considerably reduced. When attempts were made to produce powder magnets from the remaining ones, most of them were disappointing, with coercivities far below the theoretical maximum. One material stood out as being far better than the others: the samarium-cobalt alloy $SmCo_5$. The coercivity of this alloy was found to increase with decreasing particle size, in contrast to other alloys whose coercivity passed through a maximum and then decreased again. During the early 1970s, a number of different methods were developed to produce magnets from $SmCo_5$.

The rare earths are not particularly rare, but they occur in mixtures with each other, and because of their similar chemical properties, they are difficult to separate. However, by the 1970s, separation techniques had improved sufficiently to make the production of $SmCo_5$ economically worthwhile. The main problem in handling $SmCo_5$ is its chemical instability, especially in powder form. However, the problems have been solved well enough for samarium-cobalt to become an important permanent magnet material produced by many manufacturers in a number of countries.

$SmCo_5$ is hexagonal with a large uniaxial anisotropy, a fairly high Curie temperature ($\approx 720°C$) and a saturation magnetization of $\approx 770\,kA\,m^{-1}$. The theoretical maximum coercivity is at least $16\,MA\,m^{-1}$. The coercivity actually achieved is $_BH_c \approx 760\,kA\,m^{-1}$, $_MH_c \approx 3\,MA\,m^{-1}$—small compared with the maximum, but still very useful! Values of U_{max} of $\approx 200\,kJ\,m^{-3}$ have been achieved. The remanence is $\approx 780\,kA\,m^{-1}$.

The main steps in the manufacture of $SmCo_5$ magnets are powdering and sintering. Various methods have been developed to prevent the alloy from reacting chemically with its surroundings, and to prevent the formation of other intermetallic compounds, which have harmful effects on the magnetic properties. $SmCo_5$ magnets can also be produced in a polymer-bonded form. Another possibility that has been investigated is the replacement of some of the cobalt with copper. In copper-substituted alloys, non-magnetic precipitates can be formed, which increase the coercivity of the bulk material, probably by domain wall pinning. However, this possibility has not been widely exploited.

A great deal of research has been carried on the possibility of replacing samarium with one of the more abundant rare earths, or even with 'mischmetal'. Mischmetal (German for 'mixed metal' is a naturally occurring mixture of rare earths, with cerium as its main constituent, and its use would have eliminated the need to separate the mixture. However, cerium-cobalt and mischmetal-cobalt alloys have not been able to compete with $SmCo_5$ because of their inferior properties. More recent developments have however been more promising. Materials with good properties are being developed, which are based on the Sm_2Co_{17} compound with additions of other elements such as zirconium and hafnium, and on the replacement of samarium with the cheaper praeseodymium. In the best alloys, values of $M_r \approx 950\,kA\,m^{-1}$, $_BH_c \approx 800\,kA\,m^{-1}$, $U_{max} \approx 260\,kJ\,m^{-3}$ have been achieved. The improvement over $SmCo_5$ is due more to an increased remanence than to an increased coercivity.

4.2.4.5 Manganese alloys

Several manganese alloys are possible candidates for permanent magnets. The intermetallic compound MnBi has a hexagonal crystal structure and a large uniaxial anisotropy, and powder magnets with good magnetic properties have been made from it. However, it is chemically not stable enough to be useful.

A more realistic possibility is the manganese-aluminium alloy with approximate composition $Mn_{1.1}Al_{0.9}$. In this alloy, a tetragonal phase with a large anisotropy exists. This phase, which has the same crystal structure as CuAu and PtCo, is metastable at room temperature, and it has to be formed by cooling at a controlled rate or by rapid quenching followed by tempering. Cooling too slowly or tempering for too long results in decomposition into non-magnetic phases. Attempts to produce powder magnets have failed, because the coercivity of the powders is too small. The reason almost certainly lies in the formation of defects during the powdering process, which act as sites for the nucleation of reverse domains. Useful magnetic properties have however been produced by mechanical working, especially extrusion. Magnets produced in this way always contain some carbon, which has the beneficial effect of slowing down the phase transformations and thus stabilising the magnetic phase. Mn-Al-C magnets with $M_r \approx 520 \, \text{kA m}^{-1}$, $H_c \approx 220 \, \text{kA m}^{-1}$, $U_{\text{max}} \approx 56 \, \text{kJ m}^{-3}$ have been produced. The material would compete successfully with hard ferrites if the production costs were lower.

A third manganese alloy with useful properties is a silver-rich alloy containing a small amount of manganese and aluminium. It has a small saturation magnetization but a large coercivity, and it has been used in specialised applications where it is important that the magnet should not distort an external magnetic field. The structure of the alloy has been studied by electron microscopy and it was found that its permanent magnet properties were due to precipitates of manganese-aluminium.

4.2.4.6 Neodymium-iron-boron alloys

The exciting new material of the 1980s is neodymium-iron-boron. It has been known for some time that large coercivities could be achieved in rapidly cooled neodymium-iron alloys. Rapid quenching is not usually a suitable method of producing hard magnetic materials, because the properties are not easily reproducible. It was found however that in ternary alloys containing a small amount of boron, good magnetic properties could be achieved in magnets prepared by the 'conventional' techniques of breaking up into powder and sintering. The technology is thus similar to that used for samarium-cobalt. The main difference is that neodymium and iron are chemically more unstable than samarium and cobalt respectively, and therefore the powder has to be protected from oxidation even more carefully. The composition of the alloy is approximately $Nd_{15}Fe_{77}B_8$. The main magnetic phase is $Nd_2Fe_{14}B$, which has a tetragonal structure and large anisotropy. At room temperature, the alloy has extremely good properties: $M_r \approx 980 \, \text{kA m}^{-1}$, $H_c \approx 960 \, \text{kA m}^{-1}$, $U_{\text{max}} \approx 290 \, \text{kJ m}^{-3}$. These properties compare favourably with those of the best samarium-cobalt alloys, and the raw materials, neodymium and iron, are cheaper than samarium and cobalt respectively. Neodymium-iron-boron would be rapidly replacing samarium-cobalt if it were not for the fact that the Curie temperature of the former is rather low, about $310°C$, and the magnetic properties deteriorate rapidly with increasing temperature. Attempts are currently being made to improve the temperature-dependence of the properties. Replacing some of the iron with cobalt increases the Curie temperature but decreases the coercivity, and is therefore not a satisfactory solution to the problem. Nevertheless, it now seems certain that materials based on the Nd-Fe-B system will have an important part to play as permanent magnets.

4.2.5 Applications of permanent magnets

This chapter would not be complete without some mention of the applications of permanent magnets. There are many ways in which permanent magnets are used, and a detailed discussion is outside the scope of this book. There is space only for a few representative examples.

The oldest practical use of permanent magnets is probably as compasses. In the ordinary needle-type design, where a long, thin magnet is used, a high-remanence, low-coercivity material that does not fracture easily is best, and they are usually made of one of the magnet steels. More up to date designs can however make use of Alnico.

There are many ways in which magnets are used for attracting or holding iron objects. In magnetic door catches and tool racks, Alnico magnets are used in combination with soft iron yokes. In the case of refrigerator and freezer doors, flexible bonded ferrites are used. This material can be made into a long thin strip magnetized at right angles to its length, and is cheap enough to be used in large quantities. It also has the advantage that it acts as an air seal. Bonded ferrites are also used in games and toys because of their light weight, low cost and chemical stability.

Magnetic chucks are ingenious devices in which the magnetic attraction can be switched on or off. The switching is done either by physically moving some of the magnets in the chuck, or by reversing the magnetization in half the magnets using a current pulse in a coil. Magnetic chucks use Alnico or ferrite magnets.

Some frictionless bearings make use of permanent magnets. Such bearings are needed (1) for use in a vacuum where liquid lubricants could not be used, (2) at very high rotation speeds, (3) where rotation must be maintained for a long time. Some designs make use of ferrites, others of Alnico. The high Curie temperature of the latter is sometimes an advantage.

A very important application for magnets is in loudspeakers. Large loudspeakers mostly use ferrite magnets, but ferrites are unsuitable in small loudspeakers and in television sets. In the latter case, the relatively large stray fields of ferrite magnets could interfere with the operation of the picture tube, and Alnico is more often used. A related type of application (i.e. interaction with a current-carrying coil) is in measuring instruments.

There are many different designs of motors and generators using permanent magnets. The material most suitable in each case depends on the details of the design. This is an important area in which rare-earth alloy magnets are coming into increasing use. Even though the materials are more expensive than similar amounts of Alnico or ferrites, the cost of the magnet is only a small fraction of the cost of the complete device. The use of more powerful magnet materials enables the device to be made much smaller, so that the total cost can be considerably reduced.

The same considerations apply to gramophone pickups, where reduced weight is of great importance. High-quality pickups usually contain rare earth alloy magnets. Another application where small size and weight are important is in small stepping motors used for example in analogue electronic watches.

There are many areas in which rare earth alloy magnets are replacing other materials. The reduction in size of the complete device can often compensate for the increased expense of using a high-quality magnet material.

4.3 Magnetic media for information storage

4.3.1 Types of magnetic memory devices

One of the functions a permanent magnet can perform is to store information. It can 'remember' the magnetic field that has been applied to it. The reason why the magnet has a memory is hysteresis: in small fields, the magnetization depends on a larger field applied at an earlier time. With several magnets, we could store more complicated pieces of information if we could reverse the magnetization of each magnet independently of the others. This is the basic idea of magnetic memory devices, but if the device is to be of practical use, it must satisfy some stringent requirements. Firstly, we must be able to store as much information as possible in a small amount of material. Secondly, we must be able to 'write' the information without needing to use very large fields or currents (memory devices should *not* be like permanent magnets in this respect). Thirdly, once the information has been written, it should not be altered or destroyed except when deliberate action is taken to replace it with new information. And finally, it should be possible to read the information, preferably without altering it.

Magnetic media for information storage can be of three types. The first type are devices consisting of separate units each of which can be magnetized in two opposite directions. The two directions of magnetization represent a 'zero' and a 'one' respectively, so that the unit stores one 'bit' (short for *binary digit*) of information. The typical example of this type of device is the ferrite core memory. The second type are powder materials, in which pieces of information are held by groups of particles. Examples are magnetic tapes and discs. The third type are devices consisting of a single, continuous piece of material, usually a thin film, in which the information is held in the domain structure. There are many ways in which this can be done in principle.

Magnetic materials are used not only as the media in which information is recorded, but also in the devices used for writing and reading the information. Many important advances have been made in the development of read and write heads in recent years.

4.3.2 Ferrite core memories

Ferrite core memories need a material with a reasonably 'square' hysteresis loop, in which the magnetization can be reversed rapidly. The latter requirement rules out metallic materials, in which the reversal is delayed by eddy currents. A 'square' hysteresis loop means that the remanence ratio, M_r/M_s, is as close to 1 as possible, and that when a reverse field is applied, the magnetization at first decreases slowly until the field reaches a critical value, but then reverses rapidly. The ratio of the magnetization just before this critical field is reached, to the remanent magnetization is called the *squareness ratio*. To avoid demagnetizing fields, memory elements are toroidal in shape, so that they can be magnetized in either sense around the toroid. To reverse the magnetization, a current is passed along a wire threaded through the toroid. In a practical memory device, it is necessary to be able to reverse the magnetization of any one toroid. The toroids are therefore arranged in a square pattern, with separate wires threading each row and each column of toroids. To reverse the magnetization of a toroid, a current pulse is sent along both wires threading it. The magnitude of the pulse is adjusted so that the magnetization is reversed only when a current flows in both the horizontal and vertical wire simultaneously. A large squareness ratio is needed so that the magnetization is unchanged when a current flows in only one of the wires threading the toroid.

It has been predicted that the properties needed for a square hysteresis loop are a cubic crystal structure with a large negative anisotropy and a small magnetostriction. As we saw in section 3.2.3, a material with a negative anisotropy has easy directions parallel to $\langle 111 \rangle$. A cubic crystal has eight $\langle 111 \rangle$ directions but only six $\langle 100 \rangle$ directions. Therefore, in a polycrystalline material with negative anisotropy, there is a better chance that an easy direction is close to the direction of the field than in one with a positive anisotropy. The magnetization therefore only rotates by small angles when the field is removed, so that M_r is not much less than M_s. The small magnetostriction is needed to ensure that internal stresses do not induce other easy directions.

Of the simple spinel ferrites, all except cobalt ferrite have negative anisotropies. A number of square-loop materials have been developed, but the ones most widely used are Mg-Mn and Mg-Mn-Zn ferrites. Magnetic annealing can also produce squareness in some ferrites, but it leads to other properties that are undesirable in memory elements.

In the 1950s and 60s, all computers had ferrite core memories. The size of the elements decreased and the number of elements in a computer increased gradually. However, threading wires through the rings is a slow operation that cannot easily be automated, and ferrite core memories were eventually replaced by semiconductor devices.

4.3.3 Particulate recording media

Magnetic tapes and discs are very widely used for recording audio and video signals and digital information. The advantage of this type of medium is that the information can easily be written and read, is fairly permanent but easy to erase or alter, and does not require processing between the write and read operations (unlike photographic film, for example).

Tapes, floppy discs, and some hard discs consist of anisotropic magnetic particles bonded to a flexible or hard non-magnetic surface. The anisotropy is usually due to the elongated shape of the particles, although there is no reason in principle why materials with crystalline anisotropy could not be used. The latter have been used for specialised applications, but their main disadvantage is that the coercivity varies more rapidly with temperature than coercivity due to shape anisotropy.

The particles are bonded in such a way that their easy axes are predominantly parallel to the surface of the substrate, and to the direction of the magnetic field. For a magnetic tape, for example, the recording field is applied parallel to the length of the tape. An alternating field therefore produces a pattern consisting of bands perpendicular to the length of the tape, magnetized alternately parallel and antiparallel to the direction of motion of the tape. This pattern generates stray fields, which in turn induce currents in the playback head.

Materials for magnetic recording must have as large a value of M_s as possible, to provide a good signal, and the coercivity of the particles must not be either too small or too large. The useful range for H_c is from about 20 to $100 \, \text{kA m}^{-1}$ for tapes and from about 20 to $50 \, \text{kA m}^{-1}$ for discs. To achieve these values, the particles must be single domain. Any manufacturing process is bound to produce particles with a range of sizes. As we saw in section 4.2.3, particles that are too small are not ferromagnetic but superparamagnetic, i.e. have zero coercivity. Particles that are too large are multi-domain, and again have small coercivities. It is important therefore to have as large a proportion of particles in the optimum size range as possible.

Four types of materials are used in powder form as magnetic media on tapes and discs: (1) the ferric oxide, maghemite (γ-Fe_2O_3), (2) chromium dioxide (CrO_2), (3) hexagonal ferrites, and (4) metals. Each of these categories consists of a group of materials, either because the basic material can be modified by additions, or because several different basic materials belong to the same group.

4.3.3.1 Ferric oxide

The most commonly used material is γ-Fe_2O_3, which can be made into particles with length in the range 100 to 700 nm and ratio of length to diameter in the range 3:1 to 10:1. The coercivity is between about 16 and 32 kA m^{-1}, and M_s is about 400 kA m^{-1}. As described in section 2.4.4.5, the material has a cubic spinel structure containing some vacancies because of the lower iron content compared with the usual spinel composition (e.g. Fe_3O_4). The coercivity values of γ-Fe_2O_3 are in the lower part of the useful range for magnetic recording. This limits the usefulness of γ-Fe_2O_3 to recording signals in the lower range of frequencies.

A great deal of research has been carried out on ways to increase the coercivity of γ-Fe_2O_3. The most promising way is to introduce cobalt into the iron oxide. Cobalt generally tends to increase the anisotropy, and increased coercivities have been achieved in iron oxides containing a small amount of cobalt. The addition of cobalt however tends to make the magnetic properties more sensitive to increases in temperature and to stresses. The problems are being overcome, and iron oxide with added cobalt is becoming more widely used on audio cassettes. Particles with a cobalt-rich surface coating are especially useful and are used even on videotapes.

4.3.3.2 Chromium dioxide

Some of the limitations of γ-Fe_2O_3 can be overcome by using chromium dioxide. Pure CrO_2 has a useful value of M_s, about 480 kA m^{-1}, but a considerably lower Curie temperature than γ-Fe_2O_3. Chromium dioxide used for magnetic recording always has additions that facilitate the manufacture and improve the magnetic, mechanical and other physical properties. The coercivity is strongly dependent on additives, and can vary from below 40 to over 80 kA m^{-1}. One of the reasons for the larger coercivity compared with γ-Fe_2O_3 is the greater uniformity of size and shape of the particles. CrO_2 is therefore particularly suitable for high-density recording. It does however have some disadvantages. It is much more expensive than iron oxide, tends to be more abrasive to heads, and is somewhat unstable chemically, causing the magnetic properties to degrade slowly over long periods of time.

4.3.3.3 Hexagonal ferrites

Hexagonal ferrites are used for more specialised applications. As we saw in section 4.2.4.1, these materials have large coercivities, and are used mainly as permanent magnets. Their anisotropy is magnetocrystalline in origin, rather than due to shape. Their large coercivity makes them unsuitable for general recording applications, since the materials conventionally used in heads do not have high enough values of M_s to be able to alter the magnetization of these ferrites. They are used for some types of master tapes, and for magnetic strips on plastic cards. The latter have to be able to withstand being demagnetized by stray fields, and the large coercivity of hexagonal ferrites is therefore advantageous.

4.3.3.4 Metal particles

Metal particles have the advantage over all oxides of higher values of M_s. In particles with shape anisotropy, this also means larger coercivity. These properties allow thinner coatings to be used, and give higher signal levels. They do however have the disadvantage of chemical instability, which means that the surface of the particles must be protected by an inert coating such as an organic polymer. Both iron and iron-cobalt alloys have been used. Typical properties of iron particles are $M_s \approx 1.4\,\mathrm{MA\,m^{-1}}$, $H_c \approx 82\,\mathrm{kA\,m^{-1}}$.

4.3.4 Continuous film recording media

A number of different schemes have been put forward for using continuous thin films as computer memories. In the 1960s, a very large amount of research was done on planar Permalloy films, and a rather smaller amount on cylindrical films plated onto non-magnetic wires. In the event, neither system could compete with ferrite cores until semiconductor memories overtook them all. There was some feeling that plated wires could have become viable if more effort had been made to develop them. There are however important applications for continuous films, not as computer memories, but as recording media in computer peripherals, and also in video and audio recording. Some of these media are based on films with easy axis perpendicular to the surface. In such films, the domain structure can consist of very small domains, enabling a high density of information to be recorded, and giving rise to relatively large stray fields. One type of device, the *bubble domain* memory, has been the subject of research and development work since the late 1960s, and has found limited applications. Other systems based on continuous films are already in use. Besides magnetic recording, advances are also being made in magnetooptic recording, and in the development of read and write heads. Large research efforts are being expended in all these areas at present, and further important developments are expected in the next few years. The following sections give a brief description of the various possible recording systems.

4.3.4.1 Bubble domain devices

A bubble domain is more correctly a cylindrical domain in a thin film. The domain is magnetized in the opposite direction to the surrounding region, both magnetizations being perpendicular to the film. In order to be able to generate bubble domains, the film must have not only a uniaxial anisotropy with easy axis perpendicular to the film, but the anisotropy constant, K, must be large enough so that the magnetization prefers to lie parallel to the easy axis, rather than in the plane of the film, which would be favoured by demagnetizing effects. The condition necessary for this is

$$Q > 1, \qquad (4.7)$$

where Q, the *quality factor* of the film, is given by

$$Q = \frac{2K}{\mu_0 M_s^2} \qquad (4.8)$$

The domain structure in a film satisfying this condition usually consists of 'strip' domains magnetized alternately 'up' and 'down' along the easy axis, separated by parallel 180° domain walls. The walls are usually not straight but make up a complicated maze pattern (Fig. 4.4(a)). When a magnetic field (called a bias field) is applied parallel to an easy direction, the domains magnetized parallel to the field expand, and those magnetized antiparallel contract. When the field reaches a suitable value, the antiparallel domains become small, isolated cylinders. These cylinders are more colloquially referred to as bubbles. Bubbles exist only between two critical values of the field. If the field is too large, the bubbles collapse and the film becomes saturated, whereas if the field is too small, they expand into strips. These two values of the field, called the *collapse field* and the *stripeout field*, are usually quite well defined and depend on the properties of the film.

In a uniform bias field, the bubbles remain stationary, but in a non-uniform field, they experience a force in the direction of decreasing bias field. (In general, the bubble does not move in that direction, but at an angle to it, because of extra forces acting on the wall surrounding the bubble, which come into operation when the bubble starts to move. These forces, and therefore the direction of motion of the bubble, depend on the magnetization distribution in the wall.)

In order to use bubbles for information storage, it is necessary to be able to create and annihilate individual bubbles. We can then associate a 'zero' or a 'one' binary digit with the presence or absence of a bubble respectively. There are two ways in which individual bubbles can be accessed. One way is to have a fixed array of bubbles, rather like a two-dimensional crystal, and apply very localised pulses of perpendicular field at different points of the film in order to annihilate or create a single bubble. The other way is to have a fixed *bubble generator* and *bubble detector* and to move rows of bubbles continually past them. Bubbles can be generated or annihilated by applying a pulse to the generator, and the detector will detect any gap in the row of bubbles. Bubbles can be kept moving continuously by several different methods. For example, a pattern of suitably shaped islands of a soft magnetic material such as Permalloy can be deposited on the surface of the bubble film. When a field is applied parallel to the plane of the film,

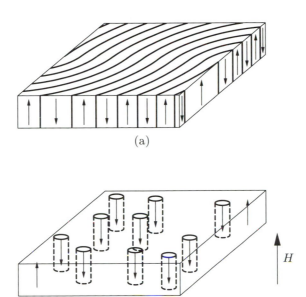

(a)

(b)

FIG. 4.4   Schematic illustration of a typical domain structure in a bubble film,
(a) in zero field, (b) in an applied field H.

the islands become magnetized and generate a non-uniform field, which is
superimposed on the bias field. The resulting field causes the bubbles to
move to particular edges of the islands. If the in-plane field is made to
rotate, the bubbles move continuously. With a suitable arrangement of the
islands, the bubbles do not just circulate round single islands, but move
from one island to another. In this way, they can be made to keep moving
past the bubble generator and detector.

For a bubble device to be useful, it must contain a large amount of information that can be rapidly written and read. The density of information
depends on the size of the bubbles, which must therefore be made as small
as possible, though not too small to be detected. Rapid access depends on
large bubble velocities. The bubble velocity is proportional to the field gradient, provided the latter is sufficiently small. If the field gradient becomes
too large, the bubbles behave in a more complicated way and their velocity
no longer increases in proportion to the field gradient. There are therefore
two important parameters that determine the velocities of the bubbles: the
ratio of velocity to field gradient, and the velocity at which the bubbles be-
gin to show a complicated behaviour. These two parameters are called the
mobility and the *critical velocity* respectively. Both must be large to be
able to achieve rapid bubble motion. Unfortunately, the bubble diameter
and the mobility both depend on the same material parameters in such a

way that optimum values of both cannot be achieved simultaneously, and reasonable compromise values must be used.

Possible materials that can be made into thin films supporting bubble domains are of four types: orthoferrites, garnets, hexagonal ferrites and amorphous metal films.

Orthoferrites were the first to be studied as possible bubble domain media. As described in section 2.4.4.4, orthoferrites are close to being antiferromagnetic. The net magnetization, M_s, which is due to the slight departure of the sublattice magnetizations from being antiparallel, is small and perpendicular to the sublattice magnetizations. Because of the small value of M_s, domains tend to be wide, leading to large diameter bubbles. This made the material convenient for studies during the early stages of the development of bubble memories, but unsuitable for practical applications.

Garnets are the only materials used as bubble media at present. They have a cubic crystal structure, and on their own, they do not possess a perpendicular anisotropy. The anisotropy can however be induced in films grown in the correct way. The usual method of fabrication is to use a single crystal of a non-magnetic garnet such as $Gd_3Ga_5O_{12}$ with a flat surface polished on it, dip it into molten magnetic garnet and keep it rotating while a thin film grows on the flat surface. The process is called liquid phase epitaxy. Films grown in this way develop a uniaxial anisotropy for two reasons. One contribution is induced by stress due to the lattice mismatch between the film and the substrate, and the other is preferential alignment of ions. Both contributions are important in producing desirable properties. All bubble garnets contain a mixture of rare earth elements. The compositions are adjusted to produce the required values of properties such as saturation magnetization M_s, and anisotropy constant K, which in turn determine the bubble parameters such as diameter and mobility. Several different mixed garnets, containing different combinations of rare earths are used, depending on the intended application. Devices with bubble sizes ranging from about $0.4\,\mu m$ to over $5\,\mu m$ have been produced.

Hexagonal ferrites have values of M_s and K both larger than those of garnets, and they can therefore support smaller bubbles. They may therefore become useful in the future. The properties may be varied by replacing some of the iron with a non-magnetic trivalent element such as aluminium or gallium. This reduces the magnetization and causes the bubble diameter to increase so that it approaches the lower values achievable in garnets.

Amorphous metal films have some advantages over oxide materials. As they do not need to be grown as single crystals, methods of preparations and substrate materials are both cheaper. A wider range of magnetic properties can be achieved than in garnets. Most amorphous bubble films contain gadolinium and cobalt or sometimes iron as the main constituents, but the properties can be varied by adding other elements. Amorphous films are not expected to be anisotropic, but uniaxial anisotropy can appear in films even when it is not expected (as in the case of garnets)! In amorphous films, the anisotropy is thought to be due to inhomogeneities induced by the method of growth. The main disadvantage of metal films is that their magnetic properties vary too rapidly with temperature in the room-temperature range. This is particularly true of films with smaller values of M_s, which support larger bubbles. The utilisation of smaller bubbles requires further advances in technology. At present, garnet films are still the most useful, but hexagonal ferrites and amorphous metal films may come into their own in the future.

From the point of view of information storage, bubble devices occupy an intermediate position between semiconductor memories, which have relatively low information densities but fast access, and magnetic discs and tapes, which have high information densities and slower access. Compared with the latter, bubble devices have the great advantage that they do not contain any moving parts. However, they are comparatively expensive, and they have not succeeded in replacing discs and tapes to a significant extent. They may find some use in applications where the system cannot easily be accessed, such as in spacecraft, and in portable computer systems.

4.3.4.2 Magnetic recording in continuous media

In bubble domain memories, the domains are kept in continuous motion, whereas the medium itself is stationary. In more conventional recording devices, the recorded pattern is in a fixed position relative to the medium, and the medium moves relative to the detector. (In the case of magnetic tapes, the tape moves and the read and write heads are stationary; in the case of discs, the disc rotates and the head moves radially.) For some applications, particulate recording media do not provide sufficiently high information densities (the number of bits that can be recorded on a unit area of film). The ultimate limit of information density in particulate media is set by the particle size, since it is necessary for each particle to be uniformly magnetized. The particle size cannot be made too small, otherwise the particles would become superparamagnetic instead of ferromagnetic (see section 4.2.1). In applications where very high information densities are needed, particulate media are being replaced by continuous films. Examples of such applications are hard discs for computers (the majority of hard discs manufactured to-day in fact have continuous films as the magnetic medium), small-format video tapes, and audio tapes for digital recording. A typical composition for the magnetic coating on a hard disc is 60% cobalt, 30% nickel and 10% chromium.

The principle of operation of these media is similar to that of particulate media: the magnetization is parallel to the plane of the film. However, the boundaries between domains magnetized in opposite directions are perpendicular to the magnetization directions of the domains, which generates strong demagnetizing fields, tending to demagnetize the films. The strength of the demagnetizing fields increases with increasing information density, which limits the density that can be achieved. Although at present the limit to the information density is set by the recording head and not by the medium, it would be useful if the density in the medium could be increased, to keep pace with possible future improvements in head design. The limitation imposed by demagnetizing effects in the medium could be overcome if the film could be magnetized in directions perpendicular to its plane: in that case, the demagnetizing fields would become weaker with increasing information density. Perpendicular recording would require more significant changes in recording technology than the change from particulate media to *in-plane* recording in continuous media, and these requirements have so far stood in the way of practical applications. A great deal of research is however being carried out to develop suitable recording media.

The main problem to be solved is that most continuous film media show a preference for being magnetized in-plane. There are a few exceptions: for example, amorphous rare-earth-transition metal films can be deposited with a perpendicular easy direction, as mentioned in section 4.3.4.1. These alloys are chemically unstable, and must therefore be protected from the atmosphere with a covering layer. Perpendicular easy directions have also been achieved in thin films of cobalt-chromium alloys.

It would be advantageous if perpendicular anisotropy could be achieved in films that do not contain rare earths. For example, cobalt would make a good perpendicular recording medium if it could be deposited with a hexagonal crystal structure, and *c*-axis perpendicular to the plane of the film, since hexagonal cobalt has a strong anisotropy with easy axis parallel to the *c*-axis. It is possible to achieve this configuration in very thin (less than 1 nm thick) cobalt films. As the thickness increases during the deposition of the film, the structure tends to become cubic, and the favourable anisotropy is lost. In order to retain the favourable anisotropy, the films would have to be made too thin to be useful, since the strength of the signal that could be read from a recorded pattern would be too weak. There is however a possible solution to this problem. Very thin cobalt layers can be interleaved with layers of a suitable non-magnetic material. In this way, the thickness of individual cobalt layers can be kept small enough for the required crystal structure to be formed, while a sufficiently large total thickness can be built up. These interleaved structures are called *multilayers* or *multilayer films*. (Multilayer structures are used in other applications too: solid-state lasers can be made from layered structures of different semiconductor compounds, and various optical devices, such as high-quality mirrors, filters and anti-reflection coatings can be made from dielectric multilayers.) Perpendicular anisotropy has been achieved in multilayers consisting of alternating layers of cobalt and any one of several non-magnetic elements (notably chromium), and these materials are currently being researched for possible application as perpendicular recording media. Moreover, using suitable interlayers, it is possible to make cobalt films retain the hexagonal crystal structure up to larger thicknesses.

4.3.4.3 Magneto-optic recording media

Magneto-optic recording is yet another method of storing information in continuous magnetic films. This method of recording combines the advantages of conventional magnetic recording and compact discs: it is capable of the same high information density as compact discs, but the recording can be altered any number of times. The reason why higher information densities can be achieved is that small areas of the film are accessed not by highly inhomogeneous magnetic fields, as produced by recording heads, but by a focused laser beam. The recording medium must be a film with a perpendicular easy axis, similar to those described at the end of the last section. Initially, the film is uniformly magnetized by a sufficiently large perpendicular magnetic field, and subsequently a smaller field is applied in the opposite direction. This field is smaller than the coercivity, and therefore it does not alter the magnetization of the film. However, if a laser beam is focused on a small area of the film, it heats the film locally. The magnetization of the heated area can then reverse because the coercivity is reduced as the temperature increases. After the laser beam is removed, this area of reversed magnetization is 'frozen in', rather like one of the cylindrical domains shown in Fig. 4.4(b). The arrangement of cylindrical domains represents the written information in a similar way to the physical depressions on a conventional compact disc. The information is also read by a laser beam, making use of the polar Kerr effect (see section 3.3.5.2).

In addition to a perpendicular easy axis, there is another requirement for magneto-optic recording: the plane of polarisation of the light must change by a sufficiently large amount when reflected from the surface of the recording medium. Some amorphous rare earth-transition metal alloy films are suitable for use as magneto-optic recording media. A typical composition for such a film is 25% terbium, 65% iron, 10% cobalt, but sometimes other rare earths such as gadolinium or dysprosium are also added. There are also possible multilayer materials: cobalt-palladium and cobalt-platinum in particular can be grown with a perpendicular easy axis, and useful magneto-optic properties.

Although magneto-optic recording has the advantage of high recording density compared with magnetic recording, it does have a disadvantage: as small areas of the film must be heated up and cooled down while the information is recorded, the process is relatively slow. Access times are therefore longer than in the case of magnetic recording. Magneto-optic recording will initially be used in cases where large amounts of information must be recorded on a semi-permanent basis.

4.3.4.4 Read and write heads

Magnetic materials are used in information storage not only as the media in which the information is recorded, but they are also important components of the heads that write and read the information in the case of magnetic recording. To end the discussion of information storage, we give a brief description of new developments in this area.

Conventional heads, such as might be found in analogue tape recorders, are miniature electromagnets, consisting of a core of soft magnetic material such as nickel-iron (see section 4.1.3.2), with a coil wound round it. For writing, the electrical signal produced by the amplifier is passed through the coil. The current passing through the coil magnetizes the core of the electromagnet, producing a magnetic field across the gap between the poles of the magnet, which in turn magnetizes the recording medium. For reading, the stray field above the surface of the medium magnetizes the read head, which induces an electromotive force in the windings. The resulting current is then processed by the amplifier circuitry.

The density of information that can be recorded depends on the size of the gap between the poles of the electromagnet. It is therefore important to keep the size of the gap to a minimum, particularly for applications in which the signal is recorded digitally. Moreover, in cases where the head as well as the recording medium has to move (such as floppy discs and hard discs), it is important to keep the mass of the head to a minimum. In these applications, the magnetic core of the head is usually in the form of a thin film of a soft magnetic material. Again, nickel-iron alloys can be used, but other possible materials are also being researched at present.

Magnetic fields can also be detected by making use of a completely different effect, which can also be used for read heads. In section 3.3.4, it was mentioned that the electrical resistance of metals can be changed by an applied magnetic field—an effect called magnetoresistance. In general, the change of resistance is quite small, at most 1–2%, but it is sufficient to be used for reading recorded information. Magnetoresistive read heads in use at present have Permalloy as the active material. The advantages of magnetoresistive read heads are that the active element can be made very small, and that the magnitude of the stray field can be measured without having to move the head relative to the medium. (For conventional heads, which work by electromagnetic induction, movement is necessary, because electromagnetic induction can only be produced by a *changing* magnetic flux.)

Note that only read heads can work by magnetoresistance—write heads must always be electromagnets of some kind. It is however possible to combine conventional write heads with a magnetoresistive element used for reading, into a single device.

Magnetoresistive heads of the future could well be based on multilayers, because it has recently been discovered that much larger changes of resistance can be produced in certain multilayers than in homogeneous alloys. In some cases, the change can be as large as 50–60%, and it has therefore been given the name *Giant magnetoresistance*. Giant magnetoresistance differs from the conventional effect not only by its size, but by its sign: in the case of conventional magnetoresistance, the resistance increases with increasing applied field, whereas in the case of giant magnetoresistance, the resistance decreases. Giant magnetoresistance has so far been found in iron-chromium multilayers, as well as in several cobalt-containing ones, notably cobalt-copper.

The mechanism responsible for giant magnetoresistance is interesting. The effect occurs in multilayers in which adjacent magnetic layers prefer to be magnetized in antiparallel directions. The antiparallel coupling is a result of a type of exchange interaction that occurs across the non-magnetic layer separating pairs of magnetic layers, and it only occurs for certain thicknesses of the non-magnetic layers. It was mentioned in section 2.1.3 that electrons have a magnetic moment called spin, which can orient itself either parallel or antiparallel to the magnetic field. In a magnetic metal, the free electrons, which carry the electric current, have their spins aligned either parallel or antiparallel to the magnetization of the metal. Electrons experience different resistance to their motion depending on the direction of their spin relative to the magnetization. When a current flows in a multilayer, electrons will move between different layers. If adjacent magnetic layers have opposite magnetization, then each electron will experience a high resistance some of the time, although at other times, it can move more freely. If a sufficiently large magnetic field is applied, it can overcome the exchange interactions between the layers, and align the magnetization of all the magnetic layers parallel to itself. Now, half the electrons will be able to move through all the layers with little resistance, whereas the other half will experience a high resistance all the time. But since half the electrons can now move more freely, the overall resistance of the multilayer decreases.

The main problem at present is that the field needed to align the magnetization of all the layers is rather large—much larger than the stray fields near the surfaces of recording media. If multilayers could be produced in which the magnetization could be switched with lower fields, then the giant magnetoresistance effect could be utilised in read heads.

Although there are many technical problems to be solved, it is likely that recording densities will continue to increase.

Appendix A

List of symbols and units

A LIST of most of the symbols used in this book, and an explanation of their meaning, are given in Table A.1. Symbols denoting scalar quantities are in italics, those denoting vector quantities are in bold. Some of the vector quantities are used in the book as if they were scalars, because we were only concerned with their magnitude, not their direction. For example, area $\mathbf{S}$ is not a scalar as implied by eq. (1.2), but a vector, whose direction is perpendicular to the area. Eq. (1.2) is really a vector equation, in which $\mathbf{m}$ and $\mathbf{S}$ are vectors and i is a scalar.

Units are given in the SI system. The meaning of the units, and some of the relationships between them, are explained after Table A.1. Dimensions are given in terms of mass (M), length (L), time (t), absolute temperature (T), and electric current (C). A quantity whose units and dimensions are omitted is dimensionless. The page numbers are those on which the meaning of the quantity is first explained.

The values of universal constants are given in Table A.2. The values are given to a larger number of decimals than are needed for the problems in appendix D, but they are still only approximate, except in the case of μ_0, which is defined to have an exact value in the SI system.

Symbol	Meaning	SI Units	Dimensions	Page
A	exchange constant	$\mathrm{J\,m^{-1}}$	MLt^{-2}	55
a	interatomic distance	m	L	55
$\mathbf{B}$	magnetic induction	T	$Mt^{-2}C^{-1}$	4
B_J	Brillouin function	—	—	141
B_s	saturation induction	T	$Mt^{-2}C^{-1}$	100
C	Curie (or Curie-Weiss) constant	K	T	10
C	Curie constant	K	T	21
C	Curie-Weiss constant (ferromagnetic)	K	T	25

Symbol	Meaning	SI Units	Dimensions	Page
C	Curie-Weiss constant (antiferromagnetic)	K	T	35
C	Curie-Weiss constant (ferrimagnetic)	K	T	45
c	constant (related to J)	—	—	21
E_a	anisotropy energy per unit volume	$J\,m^{-3}$	$ML^{-1}t^{-2}$	59
E_e	exchange energy per unit volume	$J\,m^{-3}$	$ML^{-1}t^{-2}$	54
E_H	energy per unit volume due to applied field	$J\,m^{-3}$	$ML^{-1}t^{-2}$	64
e	electron charge	C	$t\,C$	86
$\mathbf{F}$	force	N	$ML\,t^{-2}$	72
f_c	function related to B_J	—	—	22
$\mathbf{G}$	torque	N m	ML^2t^{-2}	4
g	gravitational acceleration	$m\,s^{-2}$	$L\,t^{-2}$	148
$\mathbf{H}$	magnetic field	$A\,m^{-1}$	$L^{-1}C$	3
H_c	coercivity	$A\,m^{-1}$	$L^{-1}C$	52
h	Planck's constant	J s	ML^2t^{-1}	86
i	current	A	C	3
J	total angular momentum quantum number	—	—	141
K	anisotropy constant	$J\,m^{-3}$	$ML^{-1}t^{-2}$	61
K_1	first-order anisotropy constant	$J\,m^{-3}$	$ML^{-1}t^{-2}$	59
K_2	second-order anisotropy constant	$J\,m^{-3}$	$ML^{-1}t^{-2}$	59
k	Boltzmann's constant	$J\,K^{-1}$	$ML^2t^{-2}T^{-1}$	20
L	Langevin function	—	—	141
$\mathbf{M}$	magnetization	$A\,m^{-1}$	$L^{-1}C$	5
M_r	remanence	$A\,m^{-1}$	$L^{-1}C$	52
M_s	saturation magnetization	$A\,m^{-1}$	$L^{-1}C$	10
M_s	spontaneous magnetization	$A\,m^{-1}$	$L^{-1}C$	10
$\mathbf{m}$	magnetic dipole moment	$A\,m^2$	L^2C	4
m_e	electron mass	kg	M	153
Q	quality factor	—	—	128
$\mathbf{S}$	area	m^2	L^2	4
T	absolute temperature	K	T	9
t	film thickness	m	L	87

Symbol	Meaning	SI Units	Dimensions	Page
U	energy per unit volume	$J\,m^{-3}$	$ML^{-1}t^{-2}$	110
U_{max}	maximum energy per unit volume	$J\,m^{-3}$	$ML^{-1}t^{-2}$	110
V	volume	m^3	L^3	67
$\mathbf{v}$	velocity	$m\,s^{-1}$	$L\,t^{-1}$	86
W_d	demagnetizing energy	J	ML^2t^{-2}	56
W_e	exchange potential energy	J	ML^2t^{-2}	54
w	Weiss constant	—	—	25
w_{wall}	domain wall width	m	L	60
Z	atomic number	—	—	19
α	argument of f_c, B_J and L	—	—	20
$\left.\begin{array}{c}\alpha_1\\ \alpha_2\\ \alpha_3\end{array}\right\}$	direction cosines	—	—	59
β	parameter expressing strength of exchange interaction	$J\,A^{-2}\,m^{-4}$	$ML^{-2}t^{-2}C^{-2}$	54
γ	domain wall energy per unit area	$J\,m^{-2}$	Ml^{-2}	61
γ_a	domain wall anisotropy energy per unit area	$J\,m^{-2}$	Mt^{-2}	61
γ_e	domain wall exchange energy per unit area	$J\,m^{-2}$	Mt^{-2}	61
η	fraction of ions on the B sublattice	—	—	44
θ_C	Curie temperature	K	T	11
θ_f	ferrimagnetic Curie temperature	K	T	45
θ_N	Néel temperature	K	T	11
θ_p	paramagnetic Curie temperature	K	T	45
λ	electron wavelength	m	L	86
μ	magnetic permeability	—	—	6
μ_0	permeability of vacuum	$H\,m^{-1}$	$ML\,t^{-2}C^{-2}$	4
μ_B	Bohr magneton	$A\,m^2$	L^2C	16
ξ	fraction of ions on the A sublattice	—	—	44
ρ	density	$kg\,m^{-3}$	ML^{-3}	10
Φ	magnetic flux	Wb	$ML^2t^{-2}C^{-1}$	5
χ	magnetic susceptibility	—	—	6

Symbol	Meaning	SI Units	Dimensions	Page
χ_a	initial susceptibility	—	—	51
χ_d	differential susceptibility	—	—	51
χ_m	mass susceptibility	$m^3\,kg^{-1}$	$M^{-1}L^3$	9
χ_r	reversible susceptibility	—	—	52
χ_t	total susceptibility	—	—	51

Table A.1: List of symbols and their meaning.

The following units are used in Table A.1:

A = Ampère
C = Coulomb
H = Henry
J = Joule
K = Kelvin
m = metre
N = Newton
s = second
T = Tesla
Wb = Weber

Some relationships between units:

$$1\,A = 1\,C\,s^{-1}$$
$$1\,H = 1\,J\,s^2\,C^{-2}$$
$$1\,T = 1\,Wb\,m^{-2}$$

Symbol	Meaning	Value
e	electron charge	$(1.60210 \pm 0.00007) \times 10^{-19}\,C$
h	Planck's constant	$(6.6256 \pm 0.0005) \times 10^{-34}\,J\,s$
k	Boltzmann's constant	$(1.38054 \pm 0.00018) \times 10^{-23}\,J\,K^{-1}$
m_e	electron mass	$(9.1091 \pm 0.0004) \times 10^{-31}\,kg$
μ_0	permeability of vacuum	$4\pi \times 10^{-7}\,H\,m^{-1}$
μ_B	Bohr magneton	$(9.2732 \pm 0.0006) \times 10^{-24}\,A\,m^2$

TABLE A.2   Values of universal constants.

140

Appendix B

Formulae used in the theory of paramagnetism

In chapter 2, simplified formulae were used in the discussion of paramagnetism. In this appendix, we give the full equations. The constant c, introduced in eq. (2.4), is related to the total angular momentum quantum number, J, by the equation

$$c = \frac{J+1}{3J}, \tag{B.1}$$

i.e.

$$J = \frac{1}{3c-1}. \tag{B.2}$$

J is always an integral multiple of $\frac{1}{2}$. At one extreme, when $J = \frac{1}{2}$, $c = 1$, and at the other, when $J \to \infty$, $c \to \frac{1}{3}$.

The function $f_c(\alpha)$, introduced in eq. (2.7), is actually

$$f_c(\alpha) = \frac{3c+1}{2} \coth\left(\frac{3c+1}{2}\alpha\right) - \frac{3c-1}{2} \coth\left(\frac{3c-1}{2}\alpha\right), \tag{B.3}$$

where coth stands for the hyperbolic cotangent. When written in terms of J, the function is denoted by B_J rather than f_c:

$$B_J(\alpha) = \frac{2J+1}{2J} \coth\left(\frac{2J+1}{2J}\alpha\right) - \frac{1}{2J} \coth\left(\frac{1}{2J}\alpha\right). \tag{B.4}$$

This function is called the *Brillouin function*. When $J = \frac{1}{2}$, it reduces to

$$B_{1/2}(\alpha) = \tanh\alpha, \tag{B.5}$$

where tanh is the hyperbolic tangent. When J is large, the Brillouin function approaches the *Langevin function*, $L(\alpha)$:

$$L(\alpha) = \coth\alpha - \frac{1}{\alpha}. \tag{B.6}$$

The theory of paramagnetism was established before the significance of the quantum theory became recognised, and the early version, derived by Langevin, used the function given in eq. (B.6).

Appendix C

Scalars and vectors

SOME of the quantities used in this book are vectors, as pointed out in a few places where the information was relevant. Some further explanation is included here for readers not familiar with vectors.

Vectors are quantities that have a direction as well as a magnitude. It is usual for symbols representing vector quantities to be printed in **bold** type. Examples of vectors appearing in this book are magnetic dipole moment **m**, magnetic field **H**, magnetic induction **B**, magnetization **M**, force **F**, torque **G** and velocity **v**. Quantities that only have a magnitude but not a direction, such as electric current i, magnetic flux Φ, and temperature T or θ, are scalars. They are usually printed in *italic* type. However, vector notation was only used where it was really necessary, such as in eqs (3.1), (3.2) and (3.25). Elsewhere, vector quantities were also printed in italic type. This is not a mistake—the *magnitude* of the vector is in fact a scalar, and in most cases, only the magnitude was of interest. In cases where it is particularly important to make it clear that the quantity being discussed is the magnitude of a vector, it is usual to print the vector in bold type but enclose it between vertical bars denoting 'the absolute value' of the vector. This was the case in eq. (3.1) and on the line following eq. (3.2), where the point was being made that the magnitudes of two vectors were equal.

Vectors may be represented geometrically by a straight line of length proportional to the magnitude of the vector and direction parallel to the vector. This representation may be used to illustrate the addition of two vectors. The two vectors are drawn starting from the same point, and then a line is drawn through the end of each vector parallel to the other vector. This completes a parallelogram. The diagonal passing through the common starting point is the sum of the two vectors, and the other diagonal is the difference.

Vectors may be multiplied together in two different ways. One of these ways, called *scalar multiplication*, was used in eqs (3.2) and (3.25). A scalar multiplication is denoted by a dot between the two vectors. The result of a scalar multiplication, the *scalar product*, is a scalar equal to the product of the two vectors being multiplied, times the cosine of the angle between them. The other type of multiplication is called *vector multiplication*, because its result is a vector. Vector multiplication is denoted by a multiplication sign ($\times$). The result of a vector multiplication, the *vector product*, is a vector with magnitude equal to the product of the magnitudes of the two vectors being multiplied, times the sine of the angle between them, and with direction perpendicular to both vectors. The *sense* of the vector product is determined by the *right-hand rule*: with thumb and first finger of the right hand pointing along the first and second vector being multiplied, respectively, the product is directed along the second finger. This means that if the order of the two vectors being multiplied is reversed, then the vector product reverses its direction.

Eqs (1.3), (1.4) and (3.37) actually imply vector multiplication, although they were not written in vector notation. In vector notation, they would have been

$$G = \mu_0 \mathbf{m} \times \mathbf{H}, \tag{C.1}$$

$$G = \mathbf{m} \times \mathbf{B}, \tag{C.2}$$

and

$$F = -|e|\mathbf{v} \times \mathbf{B} \tag{C.3}$$

respectively.

Appendix D

Worked examples

D.1 Questions

1. A small specimen of a paramagnetic material with susceptibility 4×10^{-4} is suspended in an inhomogeneous magnetic field of strength $10^6 \, \text{A m}^{-1}$, the vertical field gradient being $2 \times 10^7 \, \text{A m}^{-2}$. Calculate the apparent change in density of the sphere when the magnetic field is turned off.

2. Estimate the magnetic moment per atom in iron, cobalt and nickel. Express the result as the number of Bohr magnetons per atom. (Iron is body-centred cubic with lattice parameter $a = 0.286 \, \text{nm}$ and $M_s = 1.72 \, \text{MA m}^{-1}$, cobalt is close-packed hexagonal with lattice parameters $a = 0.2507 \, \text{nm}$ and $c = 0.4069 \, \text{nm}$, and $M_s = 1.43 \, \text{MA m}^{-1}$, and nickel is face-centred cubic with lattice parameter $a = 0.352 \, \text{nm}$ and $M_s = 0.484 \, \text{MA m}^{-1}$, where M_s is the saturation magnetization.)

3. The Curie temperature of the three metals in question 2 is $1044 \, \text{K}$, $1388 \, \text{K}$ and $627 \, \text{K}$ respectively. Estimate the magnitude of the Weiss constant and the Weiss molecular field for each of the metals.

4. The Curie temperature of nickel is $627 \, \text{K}$. Estimate the temperature at which the spontaneous magnetization of nickel is (a) 90%, and (b) 10% of its value at $0 \, \text{K}$.

5. At sufficiently high temperatures, the magnetic susceptibility of a certain antiferromagnetic substance is given by

$$\chi = \frac{C}{T + \theta}, \tag{D.1}$$

where $C = 0.2 \, \text{K}$, $\theta = 200 \, \text{K}$ and T is the absolute temperature. Assuming that the exchange interactions between A-sites and B-sites are much stronger than those within A-sites or within B-sites, estimate the magnetic susceptibility of a polycrystalline specimen of this substance at $0 \, \text{K}$.

6. A spherical single crystal specimen with volume V and a uniaxial anisotropy is placed in a large, uniform magnetic field at an angle θ to the easy axis of the specimen. If the anisotropy can be represented by a single constant K_1, what is the torque on the specimen? For what value of θ is the torque largest, and for what values is there no

torque? If the largest torque is $10^{-2}\,\mathrm{N\,m}$ and $V = 10^{-8}\,\mathrm{m}^{-3}$, calculate K_1.

7. The anisotropy constant of cobalt is about $5 \times 10^5\,\mathrm{J\,m}^{-3}$, and that of the $\mathrm{Nd_2Fe_{14}B}$ phase in neodymium-iron-boron is $4.5 \times 10^6\,\mathrm{J\,m}^{-3}$. Assuming that the exchange constant of both materials is $10^{-11}\,\mathrm{J\,m}^{-1}$, estimate the width and energy of domain walls in the two materials. (Both materials have a uniaxial anisotropy.)

8. A single crystal specimen of a material with uniaxial magnetocrystalline anisotropy is in the shape of a rectangular parallelepiped of length $c = 10\,\mathrm{mm}$ measured parallel to the easy direction of magnetization. The anisotropy constant is $10^5\,\mathrm{J\,m}^{-3}$, and the specimen is subdivided into domains of width $D = 5\,\mu\mathrm{m}$ by planar $180°$ walls parallel to the easy direction. Assuming that the saturation magnetization M_s is $1\,\mathrm{MA\,m}^{-1}$ and that the magnetostatic energy per unit volume is $2\mu_0 M_s^2 D/c$, calculate the exchange constant and estimate the domain wall width.

9. It can be shown that if a sphere has a uniform spontaneous magnetization M_s, then there is a uniform demagnetizing field inside it, of magnitude $\frac{1}{3} M_s$, in an opposite direction to the magnetization. Show that the demagnetizing energy of a sphere of volume V is $(\mu_0 V M_s^2)/6$.

   By comparing the demagnetizing energy of a uniformly magnetized sphere with the energy of a sphere subdivided by a plane domain wall passing through its centre, estimate the radius below which spheres of cobalt would be expected to be uniformly magnetized. Assume that the energy of the subdivided sphere is due only to the presence of the domain wall, and that the domain wall energy is the same as it would be in bulk material.

   (For cobalt, the exchange constant is $10^{-11}\,\mathrm{J\,m}^{-1}$, the anisotropy constant is $5 \times 10^5\,\mathrm{J\,m}^{-3}$, and $M_s = 1.4\,\mathrm{MA\,m}^{-1}$.)

10. A long cylindrical metal rod specimen of diameter $2\,\mathrm{mm}$ is suspended from a sensitive microbalance, with its cylinder axis vertical. A miniature furnace is placed around the specimen, and an electromagnet is placed around the furnace so that it can produce a horizontal field. When the electromagnet is turned on, the lower end of the specimen is in a magnetic field of $10^5\,\mathrm{A\,m}^{-1}$, while the field at its upper end is small in comparison. When the specimen is held at $400°\mathrm{C}$, its apparent weight changes by $4.79\,\mathrm{mg}$ when the electromagnet is switched on or off. Calculate the magnetic susceptibility of the specimen.

    When the above experiment is carried out with the specimen held at $500°\mathrm{C}$ and $600°\mathrm{C}$, the change of apparent weight is $1.42\,\mathrm{mg}$ and $0.83\,\mathrm{mg}$ respectively. What can be deduced from these results about the magnetic behaviour of the specimen?

11. An electron with energy $100\,\mathrm{keV}$ is incident normally on a thin iron specimen in an electron microscope. The thickness of the specimen is $200\,\mathrm{nm}$, and it is magnetized parallel to its surfaces. Calculate the

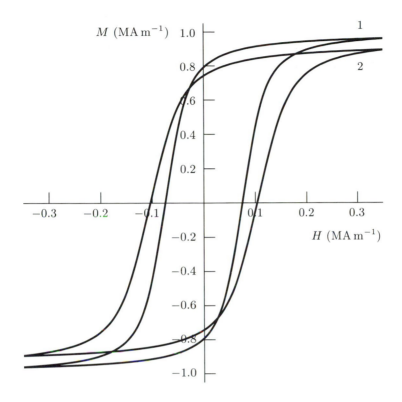

FIG. D.1   Hysteresis loops of two different materials, see question 12.

angle through which the electron is deflected after passing through the specimen. The magnetization of iron is $1.72\,\mathrm{MA\,m^{-1}}$, and the mass and charge of the electron are $9.11 \times 10^{-31}\,\mathrm{kg}$ and $1.6 \times 10^{-19}\,\mathrm{C}$ respectively.

(Note that the magnitude of the force on an electron moving perpendicular to a magnetic induction of magnitude B is given by evB, where e and v are the charge and the velocity of the electron respectively, and that the force is perpendicular to the magnetic induction and to the direction of motion of the electron.)

12. Magnets are to be made in the shape of a ring of mean radius r, containing a gap of width w. Two materials are available, their magnetization curves being shown in Fig. D.1. Estimate the range of values of the ratio w/r for which the magnetic field in the gap of a magnet made from material 1 is larger than that in the gap of a magnet made from material 2.

D.2   Answers

1. The force on a specimen of volume V and susceptibility χ is given by

$$F = \mu_0 V \chi H \frac{dH}{dz}, \tag{D.2}$$

where dH/dz is the field gradient. The increase in apparent density $\Delta\rho$ is

$$\Delta\rho = \frac{F}{gV} = \frac{\mu_0\chi}{g} H \frac{dH}{dz}, \tag{D.3}$$

where g is the gravitational acceleration ($9.81\,\mathrm{m\,s^{-1}}$). Substituting the values, we get $\Delta\rho = 1.02 \times 10^3\,\mathrm{kg\,m^{-3}}$.

2. If the volume of the unit cell is V and it contains u atoms, then the number of atoms per unit volume is u/V. Since the magnetization is the magnetic moment per unit volume, the magnetic moment per atom is $M_s V/u$. The number of Bohr magnetons per atom is this moment divided by the Bohr magneton (see section 2.1.3):

$$n = \frac{M_s V}{u\mu_B}. \tag{D.4}$$

(a) In iron, $V = a^3$ and $u = 2$, so that

$$n = \frac{a^3 M_s}{2\mu_B} \approx 2.17. \tag{D.5}$$

(b) Cobalt is hexagonal, so $V = \sqrt{3}a^2c/2$. For a close-packed hexagonal structure, $u = 2$. Hence

$$n = \frac{\sqrt{3}a^2 c M_s}{4\mu_B} \approx 1.71. \tag{D.6}$$

(c) In nickel, $V = a^3$ and $u = 4$, so that

$$n = \frac{a^3 M_s}{4\mu_B} \approx 0.569. \tag{D.7}$$

3. From eq. (2.16) with $c = 1$, the Weiss constant is given by

$$w = \frac{k\theta_C}{\mu_0 N m^2}. \tag{D.8}$$

But Nm is equal to the saturation magnetization M_s, and $m = 1\,\mu_B$. The internal field is therefore given by

$$wM_s = \frac{k\theta_C}{\mu_0\mu_B}. \tag{D.9}$$

Substituting the values, we get $1.23 \times 10^9 \, A \, m^{-1}$, $1.64 \times 10^9 \, A \, m^{-1}$, and $7.43 \times 10^8 \, A \, m^{-1}$ for iron, cobalt and nickel respectively. The Weiss constant is obtained by dividing by M_s. We get 719, 1150 and 1530 for iron, cobalt and nickel respectively.

4. The equation for spontaneous magnetization is (2.11), with $H = 0$. For nickel, we can take $c = 1$, so that eq. (B.5) applies. The spontaneous magnetization therefore satisfies the equation

$$M = Nm \tanh \left(\frac{\mu_0 mwM}{kT} \right). \tag{D.10}$$

As we saw in section 2.3.2, it is difficult to solve this equation for M at a given T, but it is easy to solve it for T at a given M. It is easily seen that the spontaneous magnetization at $0 \, K$ is equal to Nm (all magnetic moments aligned). As T approaches the Curie temperature θ_C, $M \to 0$, so the argument of the tanh function in eq. (D.10) tends to zero. The tanh function can then be replaced by its argument ($\tanh \alpha \to \alpha$). The Curie temperature is therefore

$$\theta_C = \frac{\mu_0 N m^2 w}{k}. \tag{D.11}$$

It is convenient to introduce the dimensionless variables

$$y = \frac{M}{Nm}, \tag{D.12}$$

$$x = \frac{T}{\theta_C}. \tag{D.13}$$

Substituting into eq. (D.10), we get

$$y = \tanh(y/x). \tag{D.14}$$

To solve for x, we rearrange the equation:

$$x = \frac{y}{\operatorname{arctanh} y}. \tag{D.15}$$

When the spontaneous magnetization is 90% and 10% of its value at $0 \, K$, then $y = 0.9$ and $y = 0.1$ respectively. Substituting into eq. (D.15), we get $x \approx 0.6113$ and $x \approx 0.9967$ respectively. The corresponding temperatures are obtained by multiplying by the Curie temperature, $627 \, K$. We get $383.3 \, K$ and $624.9 \, K$ respectively. The results show that magnetization decreases slowly at low temperatures (it is still 90% of its maximum value when the temperature is well over $\frac{1}{2}\theta_C$), and only varies rapidly very close to θ_C. (Many scientific calculators have the arctanh function built in, but if yours hasn't, you might find the formula

$$\text{arctanh}\, x = \tfrac{1}{2} \ln \left(\frac{1+x}{1-x} \right) \tag{D.16}$$

useful.)

5. As we can neglect interactions between magnetic moments on the same sublattice, eq. (2.36) is obeyed, and we can write

$$\chi = \frac{C}{T + \theta_N}, \tag{D.17}$$

where θ_N is the Néel temperature. At the Néel temperature, $T = \theta_N$, so the susceptibility is

$$\chi = \frac{C}{2\theta_N}. \tag{D.18}$$

At a temperature of $0\,\text{K}$, the susceptibility of a *single* crystal is zero if the field is applied parallel to the sublattice magnetizations, and is equal to the susceptibility at $T = \theta_N$ if the field is applied perpendicular. In a polycrystal, the susceptibility is the weighted average of the two values:

$$\chi = \tfrac{1}{3}\chi_{\parallel} + \tfrac{2}{3}\chi_{\perp}. \tag{D.19}$$

Since $\chi_{\parallel} = 0$, and $\chi_{\perp}$ is given by eq. (D.18), the required susceptibility is

$$\chi = \frac{C}{3\theta_N} \approx 3.33 \times 10^{-4}. \tag{D.20}$$

6. When a specimen is in a large magnetic field, the magnetization becomes aligned parallel to the field. If the easy axis is at an angle θ to this direction, the specimen has an anisotropy energy

$$W = K_1 V \sin^2 \theta. \tag{D.21}$$

(According to the question, terms with higher powers of $\sin \theta$ can be neglected, and since the specimen is spherical, the demagnetizing energy is independent of θ.) The torque G is equal to minus the differential of W with respect to θ:

$$G = -\frac{dW}{d\theta} = 2K_1 V \sin \theta \cos \theta = -K_1 V \sin 2\theta. \tag{D.22}$$

The largest torque occurs when $\sin \theta = \pm 1$, i.e. when $\theta = \pm 45°$. The torque is zero when $\sin 2\theta = 0$, i.e. when $\theta = 0$ or $90°$. (It is not hard to see why the torque is zero when $\theta = 0$: the magnetization is parallel to the easy axis, and the energy is a minimum. But when $\theta = 90°$, the magnetization is in a hard direction, and the energy is a maximum. However, the torque is still zero, because the situation corresponds to *unstable* equilibrium. The torque is largest not when the energy is largest, but when it is half-way between the largest and smallest value.)

To calculate K_1, we substitute the given values into eq. (D.22), remembering that $\sin 2\theta = 1$. We get

$$K_1 = 10^6 \,\mathrm{J\,m^{-3}}. \tag{D.23}$$

7. Since both materials have a uniaxial anisotropy, they contain mostly 180° domain walls. We can therefore use eqs (3.23) and (3.24) to calculate the width w and energy γ of domain walls. Substituting the given values gives $w = 14\,\mathrm{nm}$ and $4.7\,\mathrm{nm}$, and $\gamma = 0.014\,\mathrm{J\,m^{-2}}$ and $0.042\,\mathrm{J\,m^{-2}}$ for cobalt and $Nd_2Fe_{14}B$ respectively.

8. Let a and b be the sides of the parallelepiped perpendicular to c. The total demagnetizing energy will then be

$$W_d = \frac{2\mu_0 M_s^2 DV}{c} = 2\mu_0 M_s^2 Dab, \tag{D.24}$$

where $V = abc$ is the volume of the specimen. The only other contribution to the total energy is the energy of the domain walls. The wall energy per unit area is $2\pi\sqrt{AK}$ (see eq. (3.24). If we assume that the domain walls are perpendicular to the side a, the area of each domain wall is bc, and there are a/D domain walls in the specimen. The total energy of the specimen is therefore

$$W = 2\mu_0 M_s^2 Dab + \frac{2\pi\sqrt{AK}abc}{D}$$

$$= \left(\frac{2\mu_0 M_s^2 D}{c} + \frac{2\pi\sqrt{AK}}{D} \right) V. \tag{D.25}$$

The equilibrium value of D is given by the condition $\partial W/\partial D = 0$, which gives

$$D^2 = \frac{\pi c\sqrt{AK}}{\mu_0 M_s^2}. \tag{D.26}$$

Hence

$$A = \frac{\mu_0^2 M_s^4 D^4}{\pi^2 c^2 K}. \tag{D.27}$$

Substituting the given values (see section 1.2 for the value of μ_0), the result conveniently comes out as the round number

$$A = 10^{-11} \,\mathrm{J\,m^{-1}}. \tag{D.28}$$

The domain wall width, from eq. (3.23) is

$$w = 10^{-8}\pi \approx 3.14 \times 10^{-8} \,\mathrm{m}. \tag{D.29}$$

9. As the magnetization and the field inside the sphere are both uniform and in the opposite direction, the demagnetizing energy is given by $\frac{1}{2}\mu_0$ times the magnetization times the field times the volume of the sphere, which is $(\mu_0 V M_s^2)/6$, as required. For a sphere of radius r, the energy is $2\pi\mu_0 M_s^2 r^3/9$.

The domain wall energy per unit area can be taken to be $2\pi\sqrt{AK}$ (see eq. (3.24)). If a sphere of radius r is subdivided into two hemi-spheres by a domain wall, the area of the wall is πr^2. Therefore, the demagnetizing energy for a uniformly magnetized sphere is equal to the energy of the domain wall when

$$\frac{2\pi\mu_0 M_s^2 r^3}{9} = 2\pi^2 r^2 \sqrt{AK}, \tag{D.30}$$

giving

$$r = \frac{9\pi\sqrt{AK}}{\mu_0 M_s^2}. \tag{D.31}$$

Substituting the values, we get $r \approx 25.7$ nm.

Note that this calculation is rather unrealistic. Firstly, a sphere subdivided into two domains still has a significant demagnetizing energy, although it is much harder to calculate it than the energy of a uniformly magnetized sphere. Secondly, a domain wall in a small sphere is not likely to have the same shape, width and energy as a wall in a large specimen. For example, using eq. (3.23), the domain wall width is about 14 nm, which is more than half the radius of the particle!

10. The force on the rod is given by eq. (3.36). If the apparent change of weight is w, then the susceptibility is given by

$$\chi = \frac{2wg}{\mu_0 AH^2}, \tag{D.32}$$

where g is the gravitational acceleration, A is the cross-sectional area and H is the field at the lower end of the rod. Substituting the values, we get $\chi \approx 7.58 \times 10^{-4}$.

To determine the magnetic behaviour of the material, we plot χ^{-1} against T (see Fig. 2.8). As χ is proportional to w, we can just plot w^{-1} against T. We find that the three points lie quite close to a straight line, which passes through $w^{-1} = 0$ at $T \approx 358°\text{C} \approx 631$ K. The material is therefore ferromagnetic with a Curie temperature of 631 K.

11. When the electron enters the magnetized specimen, it experiences a force that is perpendicular to its motion and to the magnetic induc-tion. This force makes the electron move along a circular trajectory inside the specimen. However, the radius of the circle is very large compared with the thickness of the specimen, so that the electron will

only be deflected by a small angle. The deflection can therefore be calculated to a good approximation by dividing the momentum gained by the electron perpendicular to its original direction of motion by its forward momentum.

To calculate the momentum, we consider the kinetic energy, which is given by

$$\tfrac{1}{2}m_e v^2 = eV, \tag{D.33}$$

where m_e, v and e are the mass, speed and charge of the electron, and V is the potential difference the electron has been accelerated by. Hence

$$v = \sqrt{\frac{2eV}{m_e}}. \tag{D.34}$$

If the thickness of the specimen is T, the time spent by the electron inside the specimen is

$$t = \frac{T}{v}. \tag{D.35}$$

The perpendicular momentum gained by the specimen is equal to the force times the time during which the force acts. Denoting this perpendicular momentum by p, we get

$$p = evBt = BeT. \tag{D.36}$$

Hence, the angle of deflection, β, is

$$\beta = \frac{p}{m_e v} = \frac{BeT}{m_e v}. \tag{D.37}$$

Substituting for v from eq. (D.34), we get

$$\beta = BT\sqrt{\frac{e}{2m_e V}}. \tag{D.38}$$

Substituting $B = 1.72 \times 10^6$, $T = 2 \times 10^{-7}$, $e = 1.6 \times 10^{-19}$, $m_e = 9.11 \times 10^{-31}$ and $V = 10^5$, we get

$$\beta \approx 4.05 \times 10^{-4}\,\text{radians} \approx 0.023°. \tag{D.39}$$

We see that β is indeed very small, so that the approximation we made is justified. Taking into account the exact circular shape of the trajectory would have made the calculation more difficult, but it would not have made a significant difference to the result. However, there is a more serious problem with this calculation, which is that we have neglected the effect of relativity. The correct value of β given by a full relativistic calculation would have been about 5% smaller for an electron with energy 100 keV.

12.   Material 1 produces a higher field when the gap is small, but as the gap gets larger, the field produced by material 1 decreases faster than that produced by material 2 (see section 4.2.1 and Fig. 4.3). The point at which the two materials produce the same field is determined by the intersection of the two magnetization curves. By measuring the graph, we can determine that the intersection is at $H \approx -0.0264\,\mathrm{MA\,m^{-1}}$, $M \approx 0.670\,\mathrm{MA\,m^{-1}}$.

   The calculation requires some knowledge of electromagnetism. Two basic principles are needed. The first one is Ampère's law, which states that if we go round a closed path and add up the magnetic field pointing along the path multiplied by the distance of each step, the result is equal to the current enclosed. In our case, the current enclosed is zero, because we are dealing with a permanent magnet. The second principle is that the component of the magnetic induction perpendicular to any surface is the same on the two sides of the surface.

   In the present case, we apply Ampère's law to the path that goes round the ring from one poleface to the other, and then across the gap, back to the starting point. If the field in the gap is denoted by H_g, then Ampère's law gives

$$H(2\pi r - w) + H_g w = 0, \tag{D.40}$$

since the length of the path inside the magnet, where the field is H, is $2\pi r - w$, and the length of the path across the gap is w. The principle involving the magnetic induction is applied to one of the polefaces, which gives

$$\mu_0(H + M) = \mu_0 H_g, \tag{D.41}$$

or

$$H + M = H_g. \tag{D.42}$$

Note that in these equations, H, M and H_g are taken to be positive when they all point in the same sense around the circle. Eliminating H_g between eqs (D.40) and (D.42) and rearranging, we get

$$\frac{M}{H} = -\frac{2\pi r}{w}. \tag{D.43}$$

We can equate this ratio to the ratio obtained from the graph to find the value of w/r for which the two materials produce the same field. We get

$$\frac{w}{r} = \frac{2\pi \times 0.0264}{0.67} \approx 0.248. \tag{D.44}$$

Therefore, if we need a magnet with $w/r < 0.248$, we should use material 1, otherwise we should use material 2.

Appendix E

Test questions

1. Explain what is meant by (a) diamagnetism, (b) paramagnetism, (c) ferromagnetism, (d) antiferromagnetism, and (e) ferrimagnetism. Supposing you had a solid specimen with unknown magnetic properties, discuss the experiments you would need to carry out to determine to which of the above five classes the material belonged.

2. A substance contains atoms with total orbital angular momentum zero and total spin angular momentum quantum number $S = \frac{1}{2}$. Assuming that the magnetic moments of different atoms do not interact with each other, show that the magnetization M for N atoms per unit volume at temperature T in a magnetic field H is

$$M = N\mu_B \tanh\left(\frac{\mu_0 \mu_B H}{kT}\right), \tag{E.1}$$

where μ_B is the Bohr magneton. How does the magnetic susceptibility vary with temperature for small values of $\mu_0 \mu_B H / kT$?

Estimate the magnetic field that would have to be applied to such a substance at $T = 300\,\mathrm{K}$ in order to produce a magnetization equal to 3% of the saturation value. What fraction of the saturation value would be produced by the same field at $4.2\,\mathrm{K}$?

(Note:

$$\tanh x = \frac{e^x - e^{-x}}{e^x + e^{-x}}, \tag{E.2}$$

$$\tanh^{-1} x = \frac{1}{2}\ln\left(\frac{1+x}{1-x}\right), \tag{E.3}$$

$$1\mu_B = 9.27 \times 10^{-24}\,\mathrm{A\,m^2}. \tag{E.4}$$

Use $J = \frac{1}{2}$, since $S = \frac{1}{2}$ and the orbital angular momentum is zero.)

3. Explain how the Weiss theory accounts for the fact that at low temperatures ferromagnetic materials have a spontaneous magnetization. State clearly the assumptions on which the theory is based.

Show how the theory predicts a decrease of spontaneous magnetization with increasing temperature, and derive an expression for the temperature at which the spontaneous magnetization becomes zero.

Discuss briefly what other assumptions must be made in order to explain magnetic hysteresis.

4. A solid contains N atoms per unit volume, each of which carries a spin magnetic moment of $1\mu_B$. The moments are acted on by an effective internal magnetic field of magnitude wM, parallel to the magnetization M, where w is a constant. Show that in an applied magnetic field H, the magnetization satisfies the equation

$$y = \tanh\left(\frac{h+y}{t}\right), \tag{E.5}$$

where

$$y = \frac{M(T)}{M(0)}, \tag{E.6}$$

$$h = \frac{H}{N\mu_B w}, \tag{E.7}$$

$$t = \frac{T}{\theta_C}, \tag{E.8}$$

$$\theta_C = \frac{\mu_0 N \mu_B^2 w}{k}, \tag{E.9}$$

$M(T)$ is the magnetization at absolute temperature T, and k is Boltzmann's constant. What is the significance of the parameter θ_C?

The magnetization of iron, measured in a small applied field ($h \ll y$), is $1.7\,\mathrm{MA\,m^{-1}}$ at very low temperatures, and $1.2\,\mathrm{MA\,m^{-1}}$ at 835 K. Estimate the magnetic moment per atom in Bohr magnetons, the value of θ_C, and the effective internal magnetic field at low temperatures. (Iron is body-centred cubic with a lattice parameter of 0.286 nm.)

Discuss how well this model describes the properties of ferromagnetic transition elements.

(Note: see question 2 for definitions of $\tanh x$ and $\tanh^{-1} x$; take J to be the same as in question 2.)

5. Outline Néel's theory of ferrimagnetism in the case of a material having atoms on two types of site within the lattice, assuming that the exchange interactions between magnetic moments on the same type of site can be neglected in comparison with the exchange interactions between moments on different types of site. Discuss the variation of the magnetic susceptibility with temperature above the ferrimagnetic Curie temperature.

Discuss the origin of exchange interactions in spinel ferrites, and explain the effect of the addition of zinc ferrite on the magnetization of other ferrites. Estimate the magnetic moment per formula unit of the mixed ferrites $Mn_{1-c}Zn_cFe_2O_4$ and $Co_{1-c}Zn_cFe_2O_4$ as a function of c, on the assumption that all manganese and cobalt ions occupy octahedral sites and all zinc ions occupy tetrahedral sites. Is this assumption justified for $Mn_{0.4}Zn_{0.6}Fe_2O_4$ and $Co_{0.4}Zn_{0.6}Fe_2O_4$, for which the measured magnetic moments per formula unit are $6.78\,\mu_B$ and $5.78\,\mu_B$ respectively?

6. The anisotropy energy E_K per unit volume of cobalt (hexagonal close-packed crystal structure) may be approximated by

$$E_K = K_1 \sin^2 \phi + K_2 \sin^4 \phi, \tag{E.10}$$

where ϕ is the angle between the direction of the magnetization and the c-axis,

$$K_1 = (10^6 - 2 \times 10^3 T)\,\mathrm{J\,m}^{-3}, \tag{E.11}$$

$$K_2 = 10^5\,\mathrm{J\,m}^{-3}, \tag{E.12}$$

and T is the absolute temperature. How would the angle between the direction of spontaneous magnetization and the c-axis be expected to vary with temperature in the range $300\,\mathrm{K}$ to $700\,\mathrm{K}$?

7. Explain why ferromagnetic materials consist of domains that are uniformly magnetized and separated from each other by regions in which the magnetization rotates gradually. Why is this structure favoured in zero applied magnetic field over any of the following:
    (a) atomic magnetic moments pointing in random directions;
    (b) all atomic moments pointing in the same direction;
    (c) magnetization direction changing slowly and gradually throughout the material;
    (d) a domain structure with the magnetization changing direction abruptly at the domain boundaries?
    Describe the way in which the domain structure changes when a magnetic field is applied, distinguishing between reversible and irreversible changes.

8. A single crystal iron specimen in the shape of a sphere of radius $1\,\mathrm{mm}$ is suspended on a torsion balance so that it is free to rotate about the [100] direction. The specimen is placed in a large uniform magnetic field, which can be applied in any direction normal to [100]. When a field is applied in a direction making an angle $22.5°$ with [010], the specimen experiences a torque $10^{-4}\,\mathrm{Nm}$. Calculate K_1, assuming that the second and higher anisotropy constants are zero.

9. Explain what is meant by the *initial magnetization curve* of a ferromagnet, and comment on the different types of susceptibility associated with this curve. Discuss the ways in which hysteresis can arise from (a) domain wall motion and (b) magnetization rotation.

10. With respect to ferromagnetic materials, what is meant by the terms *saturation magnetization*, *remanence*, *coercivity* and *magnetic permeability*? Briefly discuss the factors influencing these parameters in

different materials. Which of these parameters are important in materials intended for:

(a) electromagnet cores,

(b) transformer cores,

(c) permanent magnets?

What other properties are important in choosing materials for these applications?

11. Discuss the factors controlling magnetic domain wall motion and magnetization rotation in ferromagnetic materials placed in a changing magnetic field, and explain how these processes lead to hysteresis. How are these processes exploited in the production of (a) soft and (b) hard magnetic materials with useful properties?

12. Explain what properties are important in permanent magnet materials, and discuss the relative importance of these properties for different applications. Compare and contrast the various materials used for permanent magnets.

   A magnet is made from material of saturation magnetization M_s in the shape of a ring of average radius r with a small gap of width g in it. Calculate the maximum magnetic field that can be generated in the gap. Why is the field always smaller than this maximum value in practice?

13. A small ferromagnetic particle with uniaxial magnetic anisotropy is uniformly magnetized along one of its easy directions. A gradually increasing magnetic field is applied in the opposite direction to the magnetization. Derive an expression for the maximum value of the field in which the original direction of magnetization may be maintained. (Assume that the anisotropy energy per unit volume of the material is given by

$$E_K = K \sin^2 \phi, \qquad (E.13)$$

where K is a constant and ϕ is the angle between the magnetization direction and the easy axis.)

   Discuss, as fully as possible, practical applications of this result. Include in your answer a description of different materials in which the result has been exploited, discussing any limitations to their properties.

Appendix F

Further reading

IF this book has provided a useful introduction to the subject of magnetic materials, the reader may feel encouraged to explore the subject in greater depth, or to seek more detailed information on topics of particular interest. There are a number of excellent books available on various aspects of magnetism and magnetic materials.

The classic work on the subject is R. M. Bozorth's *Ferromagnetism*, which appeared in the 1950s. At that time, it was still possible to collect together in one volume just about everything that was known about the entire subject, both about the science of magnetism and about an enormous range of materials. Much of the information in that book is still useful, because more recent research has concentrated on other areas. However, most of the information about materials contained in the book is on transition metals and their alloys.

The basic principles of magnetism are well treated in S. Chikazumi's *Physics of Magnetism* and A. Morrish's *The Physical Principles of Magnetism*. Both books date from the 1960s, and an updated text would be useful.

Two books dealing specifically with ferrites are *Ferrites*, by J. Smit and H. P. J. Wijn and *Oxide Magnetic Materials*, by K. J. Standley—both now over a quarter of a century old, but still useful.

Several books have been published with the aim of updating the information about magnetic materials contained in Bozorth's book. Both *Magnetic Materials*, by R. S. Tebble and D. J. Craik, and *Magnetic Materials and Their Applications*, by C. Heck, contain valuable information. The most comprehensive book on this subject is *Ferromagnetic Materials*, edited by E. P. Wohlfarth and K. H. J. Buschow, in seven volumes. (The last two volumes are entitled *Handbook of Magnetic Materials*.) These volumes contain an enormous amount of information and large numbers of references to original papers.

On the two main types of practical magnetic materials, *Magnetism and Metallurgy of Soft Magnetic Materials*, by C. W. Chen, and *Permanent Magnets in Theory and Practice*, by M. McCaig and A. G. Clegg, can be recommended.

Magnetic measurement techniques are treated in H. Zijlstra's *Experimental Methods in Magnetism*, and in *Experimental Magnetism*, edited by

Further reading

G. M. Kalvius and R. S. Tebble. The application of magnetic measurements to metallurgy is discussed in *Physical Metallurgy, Techniques and Applications*, by K. W. Andrews (volume II, section 3.3). Methods of observing magnetic domains are described in *Magnetic Domains and Techniques for their Observation*, by R. Carey and E. D. Isaac.

Index